THE QUANTITY SURVEYOR'S TOOLKIT

Essential Skills for the Modern Construction Industry

Steven Smith, Ph.D.

Wisdom Publishers

ISBN: 9798334559493
Imprint: Independently published

Printed in the United States of America

To all the aspiring and practicing quantity surveyors who tirelessly work to bring precision, innovation, and sustainability to the built environment. Your dedication to the craft inspires us all.

And to my family and mentors, whose unwavering support and encouragement have been the foundation of my journey. This book is for you.

CONTENTS

INTRODUCTION

Q uantity surveyors play a vital role in ensuring projects are completed efficiently, cost-effectively, and to the highest standards. This book is designed to be a comprehensive guide for both newcomers and seasoned professionals in the field, providing essential knowledge and tools to navigate the complexities of modern construction projects.

The construction industry is undergoing significant changes, driven by advancements in technology, a growing emphasis on sustainability, and the increasing scale and complexity of projects. In this dynamic environment, quantity surveyors must be empowered with a diverse skill set that includes not only traditional cost management and contract administration but also the ability to integrate and leverage new technologies. This book aims to bridge these areas, offering readers a holistic understanding of the quantity surveyor's role and the competencies required to excel.

The book begins by exploring the foundational aspects of quantity surveying, including the key responsibilities and career opportunities within the profession. Understanding the core tools and resources available to quantity surveyors is crucial, and the book explains the latest software and methodologies that are reshaping how projects are managed and delivered. As the book progresses, the specifics of construction contracts are considered, focusing on the critical importance of understanding contract clauses, managing risk, and navigating the legal frameworks that underpin construction projects.

The technical chapters of the book are dedicated to mastering measurement and estimating skills, which are fundamental to the practice of quantity surveying. The book provides a thorough examination of measurement techniques, cost estimating methods, and the preparation of bills of quantities, emphasizing accuracy and efficiency. These sections are enriched with practical examples and case studies, offering real-world applications and insights into best practices.

Beyond technical expertise, the book highlights the importance of soft skills such as communication, negotiation, and stakeholder management. In the high-pressure environment of construction projects, quantity surveyors must be adept at managing interpersonal dynamics and resolving conflicts. Strategies for effective communication with project stakeholders and techniques for successful negotiation and dispute resolution were explored, recognizing that these skills are as critical as technical knowledge in ensuring project success.

As sustainability becomes a cornerstone of the construction industry, quantity surveyors must also be knowledgeable about sustainable practices and green building standards. The latest trends in sustainable construction are explored, including green certifications and environmental impact assessments, and the role of quantity surveyors in advocating for and implementing sustainable solutions. The book also addresses the transformative impact of digital technologies, such as Building Information Modeling and digital measurement tools, which are revolutionizing the way quantity surveyors work.

In the final chapters, the ethical and professional responsibilities of quantity surveyors are addressed, emphasizing the importance of maintaining high standards of professionalism and ethical conduct. The role of professional bodies and associations in supporting continuous professional development is discussed, along with the significance of lifelong learning in keeping pace with industry advancements.

PART 1: INTRODUCTION TO QUANTITY SURVEYING

CHAPTER 1: WHAT IS QUANTITY SURVEYING?

The Historical Evolution of Quantity Surveying

Quantity surveying has a rich history that dates back to ancient civilizations, where the practice of measuring and valuing construction works was crucial for managing resources and costs. The roots of quantity surveying can be traced to ancient Egypt, where scribes recorded the dimensions and quantities of materials for monumental construction projects, such as the pyramids. This practice was essential for ensuring that the vast quantities of stone and labor required were accounted for accurately.

In ancient Rome, the role of quantity surveyors was further refined. Roman architects and engineers, known as "architetti" and "ingeniarii," were responsible for designing and overseeing public works, including roads, aqueducts, and buildings. These professionals kept meticulous records of materials and labor, reflecting an early form of quantity surveying. The Roman emphasis on detailed documentation and standardization laid the groundwork for modern quantity surveying practices.

The medieval period saw the emergence of master builders who acted as both architects and quantity surveyors. These individuals were responsible for the design, construction, and financial management of large projects, such as cathedrals and castles. The role of the master builder required a deep understanding of construction techniques, materials, and costs, blending the technical and financial aspects of construction management. This period also saw the development of guilds and trade organizations, which established standards and practices for construction and cost estimation.

The formalization of the quantity surveying profession began in the 17th and 18th centuries, particularly in the United Kingdom. The Great Fire of London in 1666, which destroyed much of the city, necessitated extensive rebuilding efforts. This event highlighted the need for specialized professionals who could measure and value construction work accurately, leading to the emergence of the "measurer" or "surveyor" as a distinct role. By the 18th century, the term "quantity surveyor" was in common use, denoting a professional responsible for preparing detailed estimates of construction costs based on precise measurements of quantities.

The 19th century saw significant advancements in the field of quantity surveying, driven by the Industrial Revolution. The rapid expansion of urban areas and the construction of infrastructure, such as railways and bridges, required more sophisticated methods of cost estimation and financial management. The profession became more organized, with the establishment of the Royal Institution of Chartered Surveyors (RICS) in 1868, which set professional standards and provided training for quantity surveyors. The development of standardized methods of measurement, such as the Standard Method of Measurement (SMM), further professionalized the field.

In the 20th century, quantity surveying continued to evolve with the introduction of new technologies and construction methods. The post-World War II reconstruction period saw an increased demand for quantity surveyors to manage large-scale public and private projects. The profession expanded globally, with quantity surveyors playing crucial roles in international construction projects. The advent of computers and digital technologies revolutionized the field, allowing for more precise measurements, cost estimations, and project management.

Today, quantity surveying is a dynamic and evolving profession that integrates traditional skills with modern technologies. Quantity surveyors are essential members of construction project teams, providing expertise in cost management, contract administration, and risk management. The profession continues to adapt to changes in the construction industry, including the increasing emphasis on sustainability and the use of digital tools such as Building Information Modeling. The historical evolution of quantity surveying reflects its enduring importance in the successful execution of construction projects, highlighting the profession's adaptability and resilience in the face of changing demands and technologies. Table 1 outlines the historical development of quantity surveying. It presents key periods, significant events, and their impact on the evolution of the profession.

Table 1: Historical development of quantity surveying

Date/Period	Development	Significance
Ancient Egypt	Construction of the Pyramids	Early example of cost control and project management
18th Century	Emergence of formal quantity surveying practices	Standardization of measurement and cost estimation
Early 20th Century	Establishment of professional bodies	Formalization of professional standards and

	(e.g., RICS)	ethics
Late 20th Century	Introduction of digital tools	Transformation of cost estimation and project management
21st Century	Adoption of BIM and advanced software	Enhanced accuracy, collaboration, and efficiency

The Importance of Quantity Surveying in the Construction Industry

Quantity surveying plays a pivotal role in the construction industry, providing a crucial link between the conceptual and practical aspects of building projects. The importance of quantity surveying lies in its ability to ensure that construction projects are executed efficiently, economically, and in compliance with regulatory and contractual obligations. Quantity surveyors are responsible for managing the financial aspects of construction projects, from initial cost estimates and budgeting to final account settlements, making their role indispensable for project success.

One of the primary contributions of quantity surveying is cost control. Quantity surveyors provide detailed cost planning and forecasting at the early stages of a project, enabling clients and project teams to make informed decisions regarding design, materials, and construction methods. By preparing accurate cost estimates, quantity surveyors help to prevent cost overruns, which are a common challenge in construction projects. Their expertise in analyzing and optimizing project budgets ensures that resources are allocated effectively, allowing for the efficient use of funds and minimizing financial risks.

In addition to cost control, quantity surveying is essential for contract management. Quantity surveyors are involved in the preparation and administration of construction contracts, ensuring that all parties understand and adhere to the agreed terms and conditions. They play a key role in drafting contract documents, which define the scope of work, payment schedules, and the rights and responsibilities of each party. By managing contractual obligations and monitoring compliance, quantity surveyors help to avoid disputes and ensure that projects are completed on time and within budget.

Another critical aspect of quantity surveying is risk management. Construction projects are inherently complex and involve multiple stakeholders, including clients, contractors, subcontractors, and suppliers. Each project carries potential risks related to cost, time, quality, and safety. Quantity surveyors are skilled in identifying, assessing, and mitigating these risks. They use their knowledge of construction processes and market conditions to anticipate potential challenges and develop strategies to address them. This proactive approach to risk management helps to safeguard the interests of all stakeholders and enhances the likelihood of project success.

Quality assurance is another area where quantity surveyors contribute significantly. They are responsible for ensuring that the work carried out on-site meets the required standards and specifications. Quantity surveyors perform regular site inspections and audits to verify the quality of materials and workmanship. They also monitor compliance with building codes and regulations, ensuring that projects adhere to legal and safety requirements. This focus on quality not only protects the interests of clients but also helps to maintain the integrity and reputation of the construction industry as a whole.

The role of quantity surveying extends beyond the technical and financial aspects of construction projects. Quantity surveyors often act as advisors to clients and project teams, providing valuable insights and recommendations throughout the project lifecycle. Their broad understanding of construction processes, market trends, and regulatory frameworks enables them to offer strategic guidance on project feasibility, procurement methods, and value engineering. This advisory role is particularly important in complex projects where informed decision-making is crucial for achieving project objectives.

Furthermore, quantity surveying is increasingly important in the context of sustainable construction. As the industry moves towards more environmentally friendly practices, quantity surveyors are instrumental in assessing the cost implications of sustainable design and construction methods. They help to evaluate the lifecycle costs of materials and technologies, promoting the use of energy-efficient and sustainable solutions. By incorporating sustainability considerations into cost planning and decision-making, quantity surveyors contribute to the development of green buildings and infrastructure.

Quantity surveying has become a vital component of the construction industry, providing essential services that contribute to the successful delivery of projects. Quantity surveyors ensure effective cost management, contract administration, risk mitigation, and quality assurance, all of which are crucial for achieving project goals. Their expertise and strategic insights enable them to support clients and project teams in navigating the complexities of construction projects, ultimately leading to better outcomes and enhanced project value.

The Role of a Quantity Surveyor in Construction Projects

The role of a quantity surveyor in construction projects is multifaceted, encompassing a wide range of responsibilities that are crucial for the successful planning, execution, and completion of projects. Quantity surveyors act as the financial and contractual managers within the construction team, ensuring that projects are delivered on time, within budget, and to the required quality standards. Their involvement spans the entire project lifecycle, from initial feasibility studies and cost estimation to final account settlement and post-completion evaluations.

One of the primary responsibilities of a quantity surveyor is cost management. This includes preparing preliminary cost estimates during the project's feasibility stage, which helps clients determine whether the proposed project is financially viable. Quantity surveyors develop detailed cost plans and budgets that outline the expected expenses for materials, labor, equipment, and other resources. These cost plans serve as a financial framework for the project, guiding decision-making and resource allocation throughout the construction process. By providing accurate and realistic cost estimates, quantity surveyors help prevent budget overruns and ensure that the project remains financially sustainable.

In addition to cost management, quantity surveyors are integral to contract administration. They are responsible for drafting, reviewing, and negotiating construction contracts, ensuring that all terms and conditions are clear and enforceable. Quantity surveyors work closely with legal professionals to develop contracts that protect the interests of the client and outline the responsibilities of all parties involved, including contractors, subcontractors, and suppliers. They monitor contract compliance, manage variations and change orders, and handle any disputes that may arise. This oversight ensures that the project progresses smoothly and that any issues are addressed promptly and fairly.

Risk management is another critical aspect of the quantity surveyor's role. Construction projects are subject to various risks, including financial, technical, and regulatory challenges. Quantity surveyors identify potential risks early in the project and develop strategies to mitigate them. This includes conducting thorough risk assessments, analyzing market conditions, and evaluating the impact of design changes or external factors on the project's cost and timeline. By proactively managing risks, quantity surveyors help minimize disruptions and safeguard the project's success.

During the construction phase, quantity surveyors are responsible for measurement and valuation. They measure the quantities of work completed on-site and prepare interim valuations for payment applications. This process ensures that contractors and subcontractors are paid accurately and on time for the work they have completed. Quantity surveyors also perform cost control functions, monitoring expenditures against the budget and advising on cost-saving measures where necessary. They keep detailed records of all financial transactions, providing transparency and accountability in the project's financial management.

Quality assurance is another key responsibility of quantity surveyors. They work closely with project managers and quality control teams to ensure that construction work meets the specified standards and requirements. This includes conducting regular site inspections, reviewing material specifications, and verifying that construction practices comply with building codes and regulations. Quantity surveyors play a vital role in maintaining the quality and integrity of the project, ensuring that it meets the client's expectations and industry standards.

Beyond their technical and financial duties, quantity surveyors also act as advisors and consultants. They provide expert advice on a wide range of issues, including procurement strategies, project scheduling, and value engineering. Their comprehensive understanding of construction processes and market trends enables them to offer strategic insights that enhance project outcomes. Quantity surveyors often collaborate with architects, engineers, and other professionals to optimize project designs, reduce costs, and improve overall efficiency.

The role of a quantity surveyor in construction projects is essential for ensuring effective financial management, contract administration, risk mitigation, and quality assurance. Quantity surveyors bring a unique combination of technical expertise and commercial acumen to the construction team, supporting clients and project stakeholders in navigating the complexities of the construction process. Their contributions are critical to achieving project objectives and delivering successful construction projects that meet the highest standards of quality and efficiency.

Career Paths in Quantity Surveying

Quantity surveying offers a diverse range of career paths, providing opportunities in various sectors and specializations within the construction industry. The career trajectory for quantity surveyors can vary significantly based on their interests, qualifications, and the specific needs of the industry. This diversity allows professionals to tailor their careers to their strengths and aspirations, whether they prefer working on-site, in consultancy roles, or within corporate environments.

A common starting point for many quantity surveyors is working as an assistant or trainee quantity surveyor. In these roles, individuals gain practical experience under the guidance of more experienced professionals. They are typically involved in tasks such as preparing cost estimates, assisting with tender documentation, and conducting site measurements. This early experience is crucial for developing a solid foundation in quantity surveying principles and practices. Trainee quantity surveyors often work towards professional qualifications, such as membership in the Royal Institution of Chartered Surveyors or similar professional bodies, which are highly regarded in the industry.

As quantity surveyors gain experience and expertise, they may progress to more senior roles, such as project quantity surveyor or senior quantity surveyor. In these positions, they take on greater responsibilities, including leading the cost management of projects, negotiating contracts, and managing client relationships. Senior quantity surveyors often specialize in specific areas such as cost planning, procurement, or risk management, becoming experts in their chosen fields. This specialization can lead to opportunities in niche markets, such as infrastructure projects, real estate development, or public sector construction.

Quantity surveying also offers pathways into consultancy roles. Consultancy firms provide specialized services to clients, including cost management, project management, and advisory services. Working in consultancy allows quantity surveyors to engage with a wide range of projects and clients, offering diverse and challenging work. Consultants often provide strategic advice on project feasibility, cost control, and value engineering, helping clients make informed decisions. This role requires strong analytical skills, excellent communication abilities, and a deep understanding of market trends and construction technologies.

In addition to consultancy, quantity surveyors can pursue careers in contracting firms. In these roles, they are involved in the financial and contractual management of construction projects from the contractor's perspective. This includes tasks such as preparing and submitting tender bids, managing subcontracts, and handling progress payments. Working for a contractor provides valuable insights into the practical aspects of construction, including site operations and logistics. It also offers the opportunity to be involved in the execution phase of projects, directly influencing the outcome of the construction process.

Another career path in quantity surveying is in the public sector. Government agencies and public institutions often employ quantity surveyors to oversee public infrastructure projects, manage public assets, and ensure compliance with regulatory standards. Public sector quantity surveyors may be involved in a range of projects, from transportation and utilities to education and healthcare facilities. This role often involves working closely with architects, engineers, and policymakers to deliver projects that serve the public interest.

For those interested in academia and research, quantity surveying also offers opportunities in education and training. Quantity surveyors with significant industry experience may choose to transition into teaching roles at universities or vocational training centers, educating the next generation of professionals. This path allows individuals to contribute to the advancement of the field through research, curriculum development, and thought leadership. Academics in quantity surveying often engage in research on topics such as construction economics, sustainability, and digital technologies, contributing to the body of knowledge and best practices in the industry.

Entrepreneurial quantity surveyors may also choose to establish their own consultancy firms or construction businesses. This path offers the freedom to shape one's career and business model, potentially leading to significant financial rewards and professional recognition. However, it also requires a strong understanding of business management, marketing, and client relations, as well as the ability to navigate the challenges of running a business in the competitive construction industry.

The career paths in quantity surveying are varied and offer numerous opportunities for professional growth and specialization. Whether working in consultancy, contracting, the public sector, academia, or as an entrepreneur, quantity surveyors play a vital role in the construction industry. Their expertise in cost management, contract administration, and project planning makes them indispensable to the successful delivery of construction projects, providing a rewarding and dynamic career for those in the field. Table 2 shows career paths within quantity surveying, including job titles, key responsibilities, and required qualifications. It provides a clear overview of how professionals can advance in their careers.

Table 2: Career paths within quantity surveying

Career Path	Key Responsibilities	Typical Qualifications
Trainee Quantity Surveyor	Assisting with cost estimates and measurements	Relevant degree and/or certification
Senior Quantity Surveyor	Leading cost management, client negotiations	Extensive experience, professional membership
Consultant Quantity Surveyor	Providing strategic advice, managing client relationships	Specialized expertise, professional accreditation
Contractor's Quantity Surveyor	Overseeing project finances, managing subcontracts	Experience in construction and contracting
Academic/Researcher	Teaching, conducting research, curriculum development	Advanced degrees, research publications

The Quantity Surveyor's Toolkit
(Software and Resources)

In the modern construction industry, quantity surveyors rely heavily on a range of specialized software and resources to perform their duties efficiently and accurately. The evolution of digital technologies has transformed the field, providing tools that streamline processes, enhance precision, and facilitate communication among project stakeholders. The "toolkit" for a quantity surveyor includes a variety of software applications and resources that cater to different aspects of their work, from cost estimation and measurement to project management and collaboration.

One of the fundamental tools in the quantity surveyor's toolkit is cost estimating software. These applications enable QS professionals to create detailed and accurate cost estimates for construction projects. Estimating software such as CostX, WinEst, and Sage Estimating allows quantity surveyors to input data related to materials, labor, and other expenses, and automatically calculate costs based on current market rates and project specifications. These tools often include databases of standard costs and historical data, which help in benchmarking and comparing project costs. The use of such software not only improves the accuracy of estimates but also speeds up the estimation process, allowing for quicker decision-making and more efficient project planning.

Another critical tool for quantity surveyors is Building Information Modeling (BIM) software. BIM is a digital representation of the physical and functional characteristics of a building, providing a comprehensive model that includes detailed information about every element of the construction project. Software such as Autodesk Revit, ArchiCAD, and Tekla Structures are widely used in the industry. BIM allows quantity surveyors to extract precise quantities from the digital model, ensuring that measurements are accurate and consistent. This capability is particularly valuable in complex projects where traditional measurement methods may be challenging. BIM also facilitates better coordination among project teams, as the model can be shared and updated in real time, reducing the risk of errors and omissions.

For measurement and quantification, digital takeoff software is indispensable. Tools like Bluebeam Revu, PlanSwift, and On-Screen Takeoff (OST) enable quantity surveyors to measure quantities directly from digital drawings and blueprints. These applications provide features such as digital rulers, area and volume calculators, and annotation tools, which make it easier to perform accurate takeoffs. The data obtained from digital takeoff software can be integrated with cost estimating tools, creating a seamless workflow from measurement to cost estimation. This integration helps in maintaining consistency and accuracy throughout the project lifecycle.

Project management software is another essential component of the quantity surveyor's toolkit. Applications like Microsoft Project, Primavera P6, and Asta Powerproject provide comprehensive project scheduling, resource allocation, and tracking capabilities. These tools help quantity surveyors plan and monitor project timelines, allocate resources efficiently, and track progress against the schedule. They also facilitate the management of project budgets and cash flow, providing real-time data on costs and expenditures. Effective use of project management software helps quantity surveyors ensure that projects are delivered on time and within budget, while also allowing for the efficient management of any changes or variations that arise during the construction process.

In addition to software, quantity surveyors also rely on various resources, such as industry standards, guidelines, and professional publications. Standards like the New Rules of Measurement (NRM), published by the Royal Institution of Chartered Surveyors, provide comprehensive guidance on measurement practices and procedures. These standards ensure consistency and accuracy in the quantification of construction work and are widely used in the preparation of bills of quantities and cost estimates. Professional publications, journals, and industry reports provide valuable insights into market trends, construction methodologies, and emerging technologies. Keeping abreast of these resources is crucial for quantity surveyors to maintain their knowledge and stay updated on best practices.

Collaboration and communication tools are also vital in the quantity surveyor's toolkit. Platforms like Microsoft Teams, Slack, and Zoom facilitate effective communication among project teams, clients, and other stakeholders. These tools enable real-time collaboration, document sharing, and virtual meetings, which are especially important in large or geographically dispersed projects. Effective communication is key to ensuring that all parties are aligned and that project information is accurately conveyed and understood.

The quantity surveyor's toolkit comprises a diverse array of software and resources that support the various aspects of their work. From cost estimating and measurement to project management and communication, these tools enable quantity surveyors to perform their roles with precision, efficiency, and professionalism. As technology continues to evolve, the toolkit available to quantity surveyors will expand, offering even more sophisticated and integrated solutions to meet the demands of the construction industry.

CHAPTER 2: UNDERSTANDING CONSTRUCTION CONTRACTS

Types of Construction Contracts

Construction contracts are fundamental documents that define the relationship between parties involved in a construction project. They outline the scope of work, terms and conditions, and responsibilities of each party, establishing a clear framework for project execution. There are several types of construction contracts, each suited to different project requirements and risk profiles. Understanding these types is crucial for effective project management and ensuring successful project outcomes.

One of the most common types of construction contracts is the lump sum contract, also known as a fixed-price contract. In this arrangement, the contractor agrees to complete the project for a specified sum of money. The price is agreed upon before the work begins, and it includes all costs associated with the project, such as labor, materials, and overhead. Lump sum contracts are advantageous for clients who seek a predictable budget, as the total cost is fixed regardless of the actual expenses incurred.

However, this type of contract places the risk of cost overruns on the contractor, who must absorb any additional expenses that arise during the project.

Another prevalent type is the cost-plus contract, where the client reimburses the contractor for the actual costs incurred, plus an additional fee, which may be a fixed amount or a percentage of the costs. This contract is beneficial when project scope is uncertain or when it is difficult to estimate the total cost accurately. The cost-plus contract provides flexibility to accommodate changes and unforeseen conditions. However, it can lead to budget overruns if not managed properly, as the final cost is not known until the project is complete. To mitigate this risk, a cost-plus contract may include a cap or ceiling, specifying a maximum amount that the client will pay.

Unit price contracts are another type where the work is divided into various units, and a price is established for each unit. This approach is particularly useful when the project scope involves repetitive tasks or quantities that are difficult to estimate accurately upfront. For example, in a road construction project, the cost may be calculated per unit of excavation or paving. Unit price contracts allow for adjustments based on the actual quantities of work performed, providing a fair basis for payment. However, they require meticulous measurement and recording to ensure accurate billing.

The design and build contract integrates both design and construction services under a single contract. In this arrangement, the contractor is responsible for both the design and the execution of the project. This can streamline the process, reduce conflicts between designers and builders, and potentially shorten the project duration. The client benefits from having a single point of contact, simplifying communication and coordination. However, this contract type may limit the client's control over design details and may require careful selection of a contractor with proven design capabilities.

Management contracting involves the client appointing a management contractor who oversees the construction process and coordinates the work of various subcontractors. The management contractor does not carry out the construction work directly but manages and supervises the project on behalf of the client. This type of contract can be advantageous for complex projects requiring a high level of coordination and flexibility. The client retains control over the design and can make changes more easily. However, the management contractor typically charges a fee for their services, and there is a risk of increased costs if the project is not managed effectively.

Construction management is a contract type where the client hires a construction manager to oversee the project and manage the construction process. Unlike management contracting, the construction manager is involved in the project from the early stages and provides input on cost estimation, scheduling, and procurement. The construction manager acts as an advisor and coordinator, ensuring that the project is completed on time and within budget. This approach allows for greater flexibility and the possibility of early contractor involvement. However, it requires careful management to avoid potential conflicts of interest and ensure that the construction manager's role is clearly defined.

Each type of construction contract has its own advantages and disadvantages, and the choice of contract type should be based on factors such as project scope, complexity, risk tolerance, and client preferences. Understanding these different contract types enables quantity surveyors and project managers to select the most appropriate contract for their projects, manage risks effectively, and achieve successful project outcomes. Table 3 categorizes and summarizes the main types of construction contracts.

Table 3: Types of construction contracts

Contract Type	Description	Key Features
Fixed-Price Contract	A pre-determined price is agreed upon for the entire project scope.	- Clear cost structure
Cost-Plus Contract	The contractor is reimbursed for all actual project costs incurred, plus a predetermined fee.	- Flexibility for scope changes
Time and Materials Contract	Payment is based on the time spent by labor and the materials used.	- Useful for projects with uncertain scope
Unit Price Contract	The contractor receives payment based on a pre-agreed price per unit of work completed (e.g., per cubic meter of concrete).	- Suitable for projects with measurable quantities
Design and Build Contract	A single entity is responsible for both the design and construction of the project.	- Streamlined process
Construction Management Contract	A professional is appointed to manage the construction process on behalf of the owner.	- Offers flexibility and expertise

Key Contract Clauses and Conditions

Construction contracts are intricate documents that specify the terms and conditions under which a project is executed. Understanding the key clauses and conditions within these contracts is essential for managing risks, ensuring compliance, and achieving successful project outcomes. Each clause addresses specific aspects of the contract and establishes the rights, responsibilities, and obligations of the parties involved.

1. Scope of Work: This clause defines the extent and boundaries of the work to be performed. It outlines the specific tasks, deliverables, and standards that must be met. The scope of work is crucial as it sets the expectations for both the client and the contractor, ensuring that all parties are aligned on what is to be delivered. A well-defined scope helps to prevent misunderstandings and disputes related to project deliverables.

2. Contract Price and Payment Terms: This section details the agreed-upon price for the project and specifies how and when payments will be made. It may include provisions for progress payments, final payment, and any adjustments to the contract price. Payment terms often include a schedule of payments linked to specific milestones or the completion of certain phases of the work. This clause ensures that the contractor is compensated appropriately for their work and that the client understands their financial obligations.

3. Time of Completion: This clause sets out the project timeline, including the start date, key milestones, and the final completion date. It may also include provisions for extensions of time in case of unforeseen delays. Clear timeframes are essential for project management, allowing both parties to track progress and manage expectations. The clause often specifies penalties or incentives related to timely or delayed completion.

4. Variations: Variations refer to changes in the scope of work that occur after the contract has been signed. This clause outlines the procedures for requesting and approving changes, including how they will affect the contract price and timeline. Variations can arise due to design modifications, unforeseen conditions, or client requests. A well-defined variation clause helps manage these changes systematically and ensures that all parties agree on the impact of variations.

5. Quality Standards: This clause establishes the quality requirements for the materials and workmanship involved in the project. It specifies the standards and codes that must be adhered to, ensuring that the completed work meets the required quality benchmarks. Quality standards are crucial for maintaining the integrity of the project and meeting client expectations.

6. Risk Management and Insurance: This section outlines the responsibilities of each party in managing risks and securing insurance coverage. It includes provisions for liability insurance, professional indemnity insurance, and other relevant types of coverage. This clause helps protect both the client and the contractor from potential financial losses due to accidents, damages, or other unforeseen events.

7. Dispute Resolution: Disputes may arise during the course of a construction project, and this clause defines the mechanisms for resolving such issues. It may include methods such as mediation, arbitration, or litigation. The dispute resolution clause provides a structured approach for addressing conflicts and ensures that both parties have a clear path for resolving disputes.

8. Termination: This clause specifies the conditions under which the contract may be terminated by either party. It outlines the procedures for terminating the contract, the consequences of termination, and any associated costs or penalties. Understanding the termination provisions is important for managing potential contract breaches and ensuring that the termination process is handled appropriately.

9. Force Majeure: Force majeure refers to unforeseen events or circumstances beyond the control of the parties that may impact the ability to perform contractual obligations. This clause outlines the types of events considered as force majeure and the procedures for addressing such situations. It provides a framework for managing disruptions caused by external factors such as natural disasters, strikes, or political instability.

10. Confidentiality: This clause addresses the need to protect sensitive information related to the project. It stipulates how confidential information should be handled, shared, and safeguarded. Confidentiality clauses are important for protecting proprietary information and maintaining trust between parties.

Each of these clauses plays a vital role in the construction contract, helping to establish clear expectations, manage risks, and ensure that the project proceeds smoothly. A thorough understanding of these key contract clauses and conditions enables quantity surveyors, project managers, and other stakeholders to effectively navigate the complexities of construction contracts and achieve successful project outcomes.

Risk Allocation in Construction Contracts

Risk allocation is a critical aspect of construction contracts, as it determines how potential risks and uncertainties associated with a project are shared between the parties involved. Proper risk allocation is essential for managing the financial and operational impacts of unforeseen events and ensuring that the project progresses smoothly. Effective risk management can help minimize disruptions, control costs, and achieve successful project outcomes.

1. Identifying Risks: The first step in risk allocation is identifying the various risks that may impact the project. These risks can be categorized into several types, including financial, operational, environmental, and legal risks. For example, financial risks may include cost overruns or fluctuations in material prices, while operational risks could involve delays due to subcontractor performance or equipment failures. Identifying these risks early in the project allows for a proactive approach to managing them.

2. Risk Allocation Strategies: Once risks are identified, they need to be allocated appropriately between the client and the contractor. There are several strategies for risk allocation, each with its own advantages and disadvantages.

- **Risk Transfer**: One common strategy is transferring risk from one party to another. For example, a contractor may agree to absorb the cost of certain risks, such as price fluctuations for materials, as part of a fixed-price contract. Risk transfer can provide cost certainty for the client but may increase the contractor's risk exposure.

- **Risk Sharing**: In some cases, risks are shared between the parties. This approach is often used in cost-plus contracts, where the client reimburses the contractor for actual costs plus a fee. Risk sharing can provide flexibility and accommodate changes, but it requires careful management to avoid budget overruns.

- **Risk Retention**: Certain risks may be retained by the party best able to manage them. For example, the client may retain the risk of changes in project scope, while the contractor retains the risk of delays caused by subcontractors. Risk retention allows each party to focus on areas where they have the greatest control and expertise.

3. Risk Mitigation Measures: Effective risk allocation involves not only assigning risks but also implementing measures to mitigate their impact. Mitigation measures can include:

- **Contingency Planning**: Developing contingency plans

for potential risks helps prepare for unexpected events. For example, a contingency fund may be established to cover unforeseen costs, or alternative suppliers may be identified to address potential supply chain disruptions.

- **Insurance**: Securing appropriate insurance coverage is a key risk mitigation measure. Insurance can provide financial protection against risks such as property damage, liability claims, and project delays. Ensuring that all parties have adequate insurance coverage helps reduce the financial impact of risks.

- **Contract Clauses**: Including specific clauses in the contract to address certain risks can help manage them effectively. For example, a force majeure clause can provide a framework for dealing with unforeseen events, while a liquidated damages clause can address delays and breaches of contract.

4. Risk Communication: Clear communication about risks and their allocation is essential for effective risk management. All parties involved in the project should be aware of their responsibilities and the strategies in place to address potential risks. Regular meetings and updates can help ensure that everyone remains informed and that risk management measures are implemented as planned.

5. Risk Monitoring and Review: Risk allocation is an ongoing process that requires regular monitoring and review. As the project progresses, new risks may emerge, and existing risks may change in significance. Continuous monitoring allows for timely adjustments to risk management strategies and ensures that risks are managed effectively throughout the project lifecycle.

Risk allocation in construction contracts involves identifying, assigning, and managing risks to ensure that the project progresses smoothly and achieves its objectives. By employing strategies such as risk transfer, sharing, and retention, and implementing mitigation measures, parties can effectively manage potential uncertainties and minimize their impact on the project. Clear communication and ongoing risk monitoring are crucial for successful risk management and project success.

Common Challenges and Pitfalls in Contract Management

Effective contract management is crucial for the success of construction projects, yet it often involves navigating a range of challenges and pitfalls. Understanding these common issues helps in developing strategies to mitigate their impact and ensure smooth project execution.

1. Ambiguous Contract Terms: One of the most prevalent challenges in contract management is ambiguity in contract terms. When contract clauses are not clearly defined, it can lead to misunderstandings and disputes between the parties involved. Ambiguity in scope, deliverables, or responsibilities can result in conflicting interpretations, which may cause delays, increased costs, and strained relationships. It is essential to ensure that all terms and conditions are explicitly detailed and understood by all parties to prevent such issues.

2. Change Orders and Variations: Changes in project scope, whether due to client requests or unforeseen circumstances, are common in construction projects. Managing these change orders and variations can be challenging, particularly if the process for approving and documenting changes is not well-defined. Disputes often arise regarding the cost implications and schedule adjustments resulting from changes. To mitigate these issues, it is crucial to have a robust procedure for handling change orders, including clear documentation and agreement on cost and schedule impacts.

3. Delays and Time Management: Delays are a frequent challenge in construction projects, often stemming from factors such as weather conditions, supply chain disruptions, or subcontractor performance issues. Effective time management is essential to minimize delays and ensure that the project stays on schedule. Contract management should include provisions for managing delays, such as liquidated damages clauses, and mechanisms for requesting extensions of time. Regular monitoring of progress and proactive management of potential delays are necessary to keep the project on track.

4. Cost Overruns and Budget Management: Staying within budget is a significant challenge in construction projects. Cost overruns can occur due to inaccurate cost estimates, unforeseen conditions, or mismanagement of resources. Effective budget management involves careful monitoring of expenses, regular financial reporting, and adherence to cost control measures. Contracts should include clear provisions for managing budget changes, including procedures for approving additional costs and handling financial claims.

5. Contractual Disputes: Disputes between parties can arise from various issues, such as disagreements over contract interpretation, performance deficiencies, or financial matters. Resolving disputes can be time-consuming and costly. To minimize the risk of disputes, it is important to have well-defined dispute resolution mechanisms in the contract, such as mediation, arbitration, or litigation. Establishing clear communication channels and addressing issues promptly can also help in resolving conflicts before they escalate. Table 4 outlines common types of disputes that may arise during contract management, along with typical resolution methods.

Table 4: Types of disputes

Dispute Type	Description	Potential Resolution Method
Contract Interpretation	Disagreements over the meaning of contract terms.	Mediation, Arbitration, or Negotiation; Clarify terms in future contracts.
Scope Changes	Disputes related to changes in project scope.	Implement a change management process with formal change orders.
Performance Issues	Concerns regarding the quality or timeliness of work.	Conduct performance reviews and enforce pre-defined quality standards.
Cost Overruns	Disagreements over additional costs incurred by the contractor.	Utilize detailed cost documentation for negotiation or arbitration.
Schedule Delays	Disputes related to project delays and their impact.	Review schedule clauses, consider adjusting timelines based on valid reasons.

6. Compliance with Legal and Regulatory Requirements: Construction contracts must comply with legal and regulatory requirements, which can vary by jurisdiction. Ensuring that the contract meets all relevant laws and regulations is essential to avoid legal issues and penalties. Contract management should include a review of legal and regulatory compliance, with provisions for addressing any changes in the legal landscape that may impact the project.

7. Documentation and Record Keeping: Proper documentation and record keeping are critical for effective contract management. Inadequate or incomplete documentation can lead to difficulties in tracking progress, managing changes, and resolving disputes. Contracts should specify requirements for maintaining records related to project performance, changes, payments, and communications. Implementing a systematic approach to documentation ensures that all relevant information is captured and readily accessible.

8. Performance Monitoring and Quality Control: Ensuring that work is performed to the required standards is a key aspect of contract management. Performance monitoring and quality control are essential to verify that work meets contractual requirements and industry standards. Contracts should include provisions for performance monitoring, quality assurance, and inspections to ensure that the work is completed to the specified standards.

9. Communication and Coordination: Effective communication and coordination among all parties involved in the project are crucial for successful contract management. Miscommunication or lack of coordination can lead to misunderstandings, delays, and conflicts. Contracts should outline communication protocols and responsibilities, and regular meetings and updates should be conducted to ensure that all parties are informed and aligned.

Managing construction contracts involves addressing a range of challenges and pitfalls, from ambiguous terms and change orders to delays and cost overruns. By understanding these common issues and implementing strategies to mitigate their impact, parties can achieve successful project outcomes and ensure that the project is completed on time, within budget, and to the required quality standards. Effective contract management requires careful planning, clear communication, and proactive management throughout the project lifecycle.

Best Practices for Avoiding Contractual Issues

Adopting best practices in contract management is essential for preventing issues and ensuring that construction projects proceed smoothly. These practices focus on clarity, communication, and proactive management to mitigate risks and avoid common pitfalls associated with construction contracts.

1. Draft Clear and Comprehensive Contracts: The foundation of effective contract management is a well-drafted contract. Ensure that the contract is comprehensive and unambiguous, with clearly defined terms and conditions. This includes a detailed scope of work, explicit performance criteria, and clear definitions of responsibilities. Avoid vague language and ensure that all provisions related to cost, time, and quality are precisely articulated. A thorough contract reduces the likelihood of misunderstandings and disputes.

2. Define Scope and Specifications Clearly: Precise definition of the project scope and specifications is crucial. A detailed scope of work outlines the deliverables, milestones, and quality standards expected from the contractor. Ensure that specifications are thorough and cover all aspects of the work. This clarity helps in setting realistic expectations and reduces the chances of scope creep, which can lead to disputes and additional costs.

3. Implement a Robust Change Management Process: Changes in project scope or requirements are common in construction projects. Establish a structured change management process to handle these changes effectively. This process should include procedures for requesting, evaluating, and approving changes, as well as mechanisms for documenting and communicating the impacts on cost and schedule. A robust change management process helps manage variations systematically and minimizes disputes related to changes.

4. Establish Clear Communication Channels: Effective communication is essential for managing contracts and avoiding issues. Set up clear communication channels among all project stakeholders, including clients, contractors, and subcontractors. Regular meetings, updates, and written correspondence should be used to ensure that everyone is informed about project developments, changes, and issues. Clear communication helps in addressing problems promptly and prevents misunderstandings.

5. Monitor Performance and Compliance: Regular monitoring of performance and compliance with contract terms is vital. Implement a system for tracking progress, quality, and adherence to contractual requirements. Conduct periodic inspections and reviews to ensure that work is being performed according to the agreed specifications. Address any deviations from the contract promptly to avoid escalation of issues.

6. Utilize Effective Risk Management Strategies: Proactive risk management is key to avoiding contractual issues. Identify potential risks early and develop mitigation strategies to address them. This includes implementing contingency plans, securing appropriate insurance coverage, and allocating risks clearly in the contract. By managing risks proactively, you can minimize their impact and prevent problems from affecting the project's success.

7. Ensure Compliance with Legal and Regulatory Requirements: Adherence to legal and regulatory requirements is essential for avoiding contractual issues. Ensure that the contract complies with all relevant laws, regulations, and industry standards. Regularly review and update the contract to reflect any changes in the legal landscape. Non-compliance can lead to legal disputes and penalties, so staying informed about regulatory requirements is crucial.

8. Maintain Accurate Documentation: Comprehensive and accurate documentation is essential for effective contract management. Keep detailed records of all contract-related activities, including changes, approvals, communications, and performance reports. Proper documentation provides a clear trail of evidence in case of disputes and helps in managing contract obligations and claims.

9. Foster Positive Relationships with Stakeholders: Building and maintaining positive relationships with all project stakeholders can help in avoiding conflicts and resolving issues amicably. Engage in collaborative problem-solving and foster a cooperative environment. Positive relationships facilitate open communication and mutual understanding, which can be beneficial in managing and mitigating contractual issues.

10. Review and Evaluate Contract Performance: Conduct regular reviews and evaluations of contract performance to identify areas for improvement. Assess how well the contract is being managed, and gather feedback from stakeholders. Use this information to refine contract management practices and address any recurring issues. Continuous improvement in contract management practices contributes to better project outcomes and reduced risk of contractual problems.

Best practices for avoiding contractual issues involve clear contract drafting, precise scope definition, structured change management, effective communication, and proactive risk management. Ensuring compliance with legal requirements, maintaining accurate documentation, and fostering positive relationships with stakeholders also play a critical role in successful contract management. By implementing these practices, parties can navigate the complexities of construction contracts more effectively and achieve successful project outcomes.

PART 2: MASTERING MEASUREMENT AND ESTIMATING

CHAPTER 3: INTRODUCTION TO MEASUREMENT TECHNIQUES

Standard Units of Measurement in Construction

Measurement in construction is fundamental to ensuring accuracy in both project planning and execution. The standard units of measurement are crucial for maintaining consistency and precision across various stages of a construction project.

In the construction industry, measurements are typically expressed in units that correspond to the nature of the work being performed. The primary units include length, area, volume, and weight. Length measurements are crucial for determining dimensions such as height, width, and depth. The most commonly used units for length are meters (m), millimeters (mm), and centimeters (cm) in the metric system, while feet (ft) and inches (in) are used in the imperial system.

Area measurements, which are essential for determining the extent of surfaces, are usually expressed in square meters (m²) or square feet (ft²). For larger projects or certain types of construction, square yards (yd²) may also be utilized. Volume measurements, necessary for quantifying materials such as concrete or earth, are typically expressed in cubic meters (m³) or cubic feet (ft³).

Weight measurements, important for understanding the load-bearing requirements and the quantities of materials, are often given in kilograms (kg) or tons (t) in the metric system, and in pounds (lb) or short tons (US tons) in the imperial system.

In addition to these fundamental units, construction projects may also use specialized units tailored to specific needs. For example, in the context of structural engineering, units such as kilonewtons (kN) might be used to express forces, while in road construction, units like the Marshall stability value may be employed for asphalt mix testing.

Consistency in the use of standard units ensures that all stakeholders in a construction project are working from the same measurements, which minimizes errors and misunderstandings. Quantity surveyors, architects, engineers, and contractors need to have a clear understanding of these units and their proper application in various contexts to achieve accurate and reliable results throughout the project lifecycle.

This understanding of standard units of measurement sets the stage for more complex aspects of construction measurement, including the techniques used for taking off quantities and the application of technology in modern measurement practices.

Taking Off Quantities from Drawings and Specifications

Taking off quantities, also known as quantity takeoff, is a critical process in construction measurement that involves extracting and quantifying the amounts of materials and work required from construction drawings and specifications. This process is essential for accurate cost estimation, resource planning, and project management.

The process of taking off quantities begins with a thorough review of the construction drawings, which typically include architectural, structural, and mechanical plans. Each type of drawing provides different details pertinent to various aspects of the construction project. Architectural drawings offer insights into the layout, dimensions, and finishes, while structural drawings detail the load-bearing elements and structural components. Mechanical and electrical drawings provide information on systems such as HVAC, plumbing, and electrical installations.

Quantity takeoff involves several steps

- **Review of Drawings**: The first step is to carefully examine the drawings to understand the scope of work and the details provided. This involves identifying and interpreting all relevant elements, including dimensions, materials, and construction methods.

- **Measurement**: Accurate measurement of the various elements depicted in the drawings is crucial. This involves using tools such as scales, digital calipers, and measuring tapes to determine dimensions. In the case of complex or large-scale projects, digital tools such as software applications may be employed to enhance precision.

- **Quantification**: After measuring, the next step is to quantify the materials and work required. This involves calculating areas, volumes, and lengths based on the measurements taken. For example, the quantity of paint required for a surface is determined by multiplying the area of the surface by the paint coverage rate.

- **Recording and Documentation**: Accurate recording of the quantities is essential. This is typically done using spreadsheets, databases, or specialized quantity takeoff software. The recorded quantities must be cross-checked against the drawings and specifications to ensure accuracy.

- **Reconciliation with Specifications**: It is important to cross-reference the quantities with the project specifications to ensure that all requirements are met. This involves checking that the quantities align with the specified materials and standards.

- **Cost Estimation**: Once the quantities are determined, they are used to estimate the costs associated with the materials and labor required. This step involves applying unit rates to the quantities to generate cost estimates.

Quantity takeoff is a thorough process that requires a high degree of accuracy and attention to detail. It is essential for ensuring that all necessary materials are accounted for and that the project budget is accurately estimated. Properly conducted quantity takeoffs contribute to effective project planning, cost control, and resource management, ultimately supporting the successful execution of construction projects.

Using Measurement Software Tools

The integration of measurement software tools into the construction industry has revolutionized the way quantity surveying and measurement tasks are performed. These digital tools offer enhanced accuracy, efficiency, and flexibility compared to traditional manual methods. Measurement software tools have become indispensable for modern quantity surveyors, enabling them to streamline their processes and handle complex projects with greater ease.

Measurement software tools typically include functionalities for digital takeoff, cost estimation, and project management. These tools often come with features such as automated calculations, integration with Building Information Modeling, and advanced reporting capabilities.

1. Digital Takeoff Software: Digital takeoff software allows users to perform quantity takeoffs directly from digital drawings and blueprints. By using electronic versions of construction drawings, users can accurately measure dimensions and quantities with precision. These tools often provide functionalities such as auto-scaling, which adjusts measurements based on the drawing scale, and data extraction capabilities, which automate the process of quantifying materials.

2. Building Information Modeling Integration: Many modern measurement software tools are integrated with BIM platforms. BIM provides a digital representation of the physical and functional characteristics of a project. Measurement tools that integrate with BIM can directly extract quantities from the BIM model, ensuring that the measurements are consistent with the most current project data. This integration facilitates real-time updates and reduces the risk of discrepancies between the model and the drawings.

3. Cost Estimation and Budgeting: Measurement software tools often include modules for cost estimation and budgeting. These modules allow users to apply unit rates to quantities extracted from the drawings to generate detailed cost estimates. The software can also track changes and variations in project scope, providing updated cost forecasts and facilitating effective budget management.

4. Advanced Reporting and Documentation: Measurement software tools typically offer advanced reporting features, enabling users to generate comprehensive reports and documentation. These reports can include detailed summaries of quantities, cost estimates, and project progress. The ability to produce professional and accurate reports is crucial for effective communication with stakeholders and for maintaining clear project records.

5. Integration with Other Project Management Tools: Many measurement software tools can integrate with other project management applications, such as scheduling and procurement tools. This integration enhances overall project coordination by providing a unified platform for managing various aspects of the project, from measurement and cost estimation to scheduling and resource allocation.

6. User Training and Support: To fully leverage the capabilities of measurement software tools, proper training and support are essential. Users must be familiar with the software's functionalities and best practices to ensure that they can effectively utilize the tools for their specific needs. Training programs and technical support services are often provided by software vendors to assist users in maximizing the benefits of the software.

The adoption of measurement software tools represents a significant advancement in the field of quantity surveying. These tools enhance accuracy, streamline workflows, and support more efficient project management. By embracing digital technologies, quantity surveyors can improve their ability to manage complex projects and deliver high-quality outcomes. Table 5 summarizes various measurement tools and techniques.

Table 5: Measurement tools and techniques

Measurement Tool/ Technique	Application	Advantages	Limitations
Manual Measurement	Basic length, area, and volume measurements	Simple, low-cost	Prone to human error, time-consuming
Digital Tape Measure	Linear distance measurement	High accuracy, portability	Limited range, requires battery power
Laser Distance Meter	Precise distance measurement	High accuracy, long range	Requires line of sight, affected by environmental conditions
Total Station	Surveying and construction site measurement	High accuracy, versatility	Expensive, requires skilled operator
Building Information Modeling (BIM)	Integrated measurement and design	Comprehensive data, real-time updates	High initial cost, requires software expertise
Geographic Information Systems (GIS)	Large-scale mapping and spatial analysis	Integrates multiple data sources	Complex, requires specialized training

The Role of Technology in Modern Measurement Techniques

Technology has profoundly transformed the field of quantity surveying and construction measurement. Advances in digital tools, software, and automated systems have significantly enhanced the accuracy, efficiency, and scope of measurement techniques. The role of technology in modern measurement techniques is multifaceted, influencing various aspects of construction projects from design through to execution.

1. Precision and Accuracy: Technology has greatly improved the precision and accuracy of measurement techniques. Digital measuring tools, such as laser distance meters and electronic total stations, provide highly accurate measurements with minimal error. These tools are capable of capturing precise dimensions and distances quickly, reducing the likelihood of manual errors that can occur with traditional measurement methods. The use of advanced technology ensures that the quantities extracted from drawings and specifications are accurate, which is crucial for cost estimation and project planning.

2. Enhanced Data Management: Modern measurement technologies facilitate effective data management through digital platforms and software applications. Measurement data can be captured, stored, and organized in digital formats, allowing for easier access, retrieval, and analysis. Software tools that integrate with Building Information Modeling and Geographic Information Systems enable the consolidation of measurement data with other project information, providing a comprehensive view of the project and supporting informed decision-making.

3. Integration with Building Information Modeling: Building Information Modeling represents a significant technological advancement in the construction industry. BIM provides a digital representation of a building's physical and functional characteristics. Measurement tools that integrate with BIM can extract quantities directly from the model, ensuring that the measurements are consistent with the most current project data. BIM integration also allows for real-time updates and adjustments, improving coordination and reducing the risk of discrepancies between the design and the construction phases.

4. Automation and Efficiency: Automation plays a key role in modern measurement techniques, streamlining processes and enhancing efficiency. Automated measurement systems, such as robotic total stations and drone surveying, can cover large areas quickly and with high precision. These technologies automate the process of capturing measurements, reducing the time and labor required for manual data collection. Automated systems also facilitate the generation of detailed reports and documentation, further increasing efficiency.

5. Real-Time Data and Collaboration: Technology enables real-time data collection and sharing, which enhances collaboration among project stakeholders. Digital measurement tools and software applications allow for the immediate transmission of measurement data to team members, contractors, and clients. This real-time communication helps to address issues promptly and ensures that all parties are working with the most up-to-date information.

6. Visualization and Analysis: Technological advancements in visualization tools, such as 3D modeling and virtual reality, provide new ways to analyze and interpret measurement data. These tools allow stakeholders to visualize the project in three dimensions, facilitating a better understanding of spatial relationships and design elements. Enhanced visualization aids in identifying potential issues and making informed decisions during the planning and execution phases.

7. Training and Adaptation: As technology continues to evolve, ongoing training and adaptation are essential for quantity surveyors and construction professionals. Staying current with technological advancements and acquiring the necessary skills to effectively use new tools and systems is crucial for maintaining competitiveness and achieving successful project outcomes.

The integration of technology into modern measurement techniques has revolutionized the field of quantity surveying. By improving precision, efficiency, and data management, technology supports better decision-making and enhances overall project performance. Leveraging technological advancements is essential for quantity surveyors to stay at the forefront of the industry and deliver high-quality results in an increasingly complex and dynamic construction environment.

Practical Examples and Case Studies

Practical examples and case studies play a crucial role in understanding the application of measurement techniques in real-world scenarios. They provide tangible insights into how measurement principles are applied in various projects and demonstrate the effectiveness of different tools and methodologies. This section will explore several practical examples and case studies that illustrate key concepts and best practices in measurement and estimating.

1. Residential Building Project: Precision in Quantity Takeoff

In a residential building project, accurate quantity takeoff is essential for budgeting and procurement. For instance, consider a project involving the construction of a three-bedroom house. The measurement process begins with analyzing architectural drawings, including floor plans and elevations. Digital measurement tools, such as software that integrates with Building Information Modeling, are used to extract quantities of materials like concrete, bricks, and timber.

A case study of this project highlights the use of digital takeoff software to measure the quantities of various building elements. The software's auto-scaling feature adjusts measurements according to the drawing scale, while its data extraction capabilities automate the process of quantifying materials. This approach ensures high accuracy and reduces the potential for human error. The case study also demonstrates how the integration of measurement software with BIM facilitates real-time updates and coordination, improving overall project efficiency.

2. Commercial Office Building: Cost Estimation and Budgeting

For a commercial office building project, cost estimation and budgeting are critical for financial planning and project management. Consider a case study of a multi-story office building where cost estimating software is utilized to calculate the overall project cost. The software applies unit rates to the quantities extracted from the digital drawings, generating detailed cost estimates for various construction elements, such as structural steel, glazing, and finishing materials.

The case study illustrates the importance of accurate cost estimation in managing project budgets and ensuring financial viability. The software's advanced reporting features provide comprehensive summaries of cost estimates, enabling stakeholders to track expenses and make informed decisions. Additionally, the case study highlights the role of measurement tools in managing cost risks by tracking changes and variations in project scope.

3. Infrastructure Project: Integration with Geographic Information Systems

In large-scale infrastructure projects, such as road construction or bridge building, the integration of measurement tools with Geographic Information Systems (GIS) offers significant benefits. A case study of a highway construction project demonstrates how GIS technology is used to capture and analyze measurement data for the project's alignment and terrain.

The case study details the use of drone surveying and automated measurement systems to gather data on the project site. The GIS platform consolidates this data with other project information, such as environmental impact assessments and design parameters. The integration of measurement tools with GIS enhances project planning and coordination by providing a comprehensive view of the project area and facilitating effective decision-making.

4. Industrial Facility: Automation and Efficiency

For an industrial facility project, automation plays a key role in improving measurement efficiency. A case study of a manufacturing plant construction project showcases the use of robotic total stations and automated measurement systems to survey large areas and capture precise data. The robotic total stations automate the measurement process, reducing the time and labor required for data collection.

The case study highlights the benefits of automation in enhancing project efficiency and accuracy. By using automated systems, the project team can generate detailed measurement data quickly and produce accurate reports. The case study also demonstrates how automation supports effective project management by providing real-time data and facilitating collaboration among stakeholders.

5. Heritage Building Restoration: Challenges and Solutions

Restoring heritage buildings presents unique challenges in measurement and estimating due to the complexity of historical structures. A case study of a heritage building restoration project illustrates how measurement techniques are adapted to address these challenges. The project involves detailed measurement of existing conditions, including structural elements and finishes, to develop a restoration plan.

The case study describes the use of advanced measurement tools, such as laser scanners and 3D modeling, to capture precise data on the building's condition. The integration of this data with restoration plans ensures that the historical integrity of the building is preserved while meeting modern construction standards. The case study highlights the importance of accurate measurement in achieving successful restoration outcomes and maintaining the building's historical value.

Practical examples and case studies provide valuable insights into the application of measurement techniques in various contexts. They demonstrate the effectiveness of different tools and methodologies, illustrating how measurement principles are applied to achieve successful project outcomes. By examining these real-world scenarios, quantity surveyors and construction professionals can gain a deeper understanding of measurement techniques and their impact on project performance.

CHAPTER 4: BUILDING COST ESTIMATING FUNDAMENTALS

Cost Estimating Techniques

C ost estimating is a critical component of construction project management, providing the financial foundation for successful project planning and execution. Accurate cost estimates are essential for ensuring that projects are completed within budget and that financial resources are allocated effectively. This section examines two fundamental cost estimating techniques: Elemental Cost Estimating and Cost in Use Estimating.

Elemental Cost Estimating

Elemental cost estimating is a method where the total cost of a project is broken down into its individual components or elements. This approach allows for a detailed analysis of various parts of the construction project, facilitating a comprehensive understanding of cost distribution and helping to identify potential cost-saving opportunities.

The elemental approach categorizes the cost into distinct elements such as substructure, superstructure, finishes, and services. Each category represents a specific aspect of the construction work, and costs are estimated separately for each element. This method is particularly useful in the early stages of project planning when detailed design information may not yet be available. By estimating costs at the elemental level, project managers can develop a preliminary budget that reflects the major components of the project.

Cost in Use Estimating

Cost in use estimating, also known as life-cycle costing, focuses on the total cost of ownership of a building or facility over its entire life span. This technique considers not only the initial construction costs but also the long-term costs associated with operation, maintenance, and disposal.

The principle behind cost in use estimating is to provide a more comprehensive view of the financial implications of a project by including ongoing expenses. These expenses can include energy costs, maintenance and repair costs, and the cost of upgrading or replacing components. By evaluating these factors, cost in use estimating helps to identify the most cost-effective solutions from a long-term perspective.

For instance, while a more expensive building material might have a higher initial cost, it could offer greater durability and lower maintenance costs over the building's lifespan. This approach aids in making informed decisions that balance initial investment with long-term financial performance.

Comparison and Application

Both elemental and cost in use estimating techniques serve different purposes and are applicable at various stages of a project. Elemental cost estimating is advantageous during the early phases when detailed project information is limited, providing a structured approach to budgeting based on known quantities and components. Cost in use estimating, on the other hand, is crucial for evaluating the overall financial impact of a project, incorporating long-term costs that affect the total cost of ownership. Table 6 presents a comparison of various cost estimating techniques, including elemental cost estimating and cost in use.

Table 6: Elemental Cost Estimating and Cost in Use

Cost Estimating Technique	Description	Advantages	Limitations
Elemental Cost Estimating	Breaks down project into elements (foundations, walls, etc.) for cost estimation.	Detailed analysis, comparison across projects	Time-consuming, requires detailed information
Cost-in-Use Estimating	Considers total cost of ownership, including operation and maintenance.	Comprehensive cost view, useful for lifecycle analysis	Requires detailed operational data, complex calculations

In practice, combining these methods can provide a more thorough financial analysis. For example, an initial elemental estimate can be refined by incorporating life-cycle cost considerations, ensuring that both immediate and future financial impacts are accounted for. This integrated approach supports more informed decision-making and promotes effective cost management throughout the project's lifecycle.

Table 7 compares different cost estimation and measurement software tools, highlighting their features, benefits, and suitability for different types of projects.

Table 7: Cost estimation and measurement software tools

Software Tool	Key Features	Benefits	Suitable For
CostX	2D and 3D takeoff, integrated cost estimating	High accuracy, BIM integration, user-friendly	Large-scale construction projects
Bluebeam Revu	PDF-based takeoff, real-time collaboration	Efficient for document-based measurements, easy to use	Small to medium-sized projects
WinQS	Detailed cost estimation, project management	Comprehensive features for cost control and project tracking	All types of construction projects
PlanSwift	Digital takeoff, cost estimation tools	Easy integration with other software, efficient takeoff process	Medium to large-scale projects
On-Screen Takeoff	Digital measurement from plans and drawings	Quick setup, suitable for various types of projects	Various types of projects
Estimator360	Cloud-based estimation, real-time updates	Accessible from multiple devices, collaborative features	Remote teams, varying project scales

Sage Estimating	Advanced cost estimation and analysis tools	Detailed reporting, integrates with accounting systems	Large projects with complex budgeting needs

Cost Estimating at Different Project Stages

Cost estimating is a dynamic process that evolves throughout the lifecycle of a construction project. Accurate cost estimation at each stage is crucial for effective financial management, ensuring that the project remains within budget and meets its financial objectives. This section outlines the cost estimating process at various project stages, highlighting the key characteristics and considerations for each phase.

1. Preliminary or Conceptual Stage

During the preliminary or conceptual stage, the project's scope and design are not yet fully defined. At this point, cost estimates are often based on broad assumptions and historical data from similar projects. The primary objective is to provide a rough estimate of the project's cost to support initial decision-making and feasibility studies.

Estimates at this stage are typically prepared using conceptual estimating techniques, such as square footage rates or unit costs. These estimates offer a preliminary budget that helps stakeholders assess the project's viability and make informed decisions about moving forward. The accuracy of these estimates is relatively low, given the limited detail available.

2. Design Development Stage

As the project progresses to the design development stage, the design becomes more detailed, and cost estimates can be refined. At this stage, the project's specifications, drawings, and design documents are more developed, allowing for a more accurate estimation of costs.

Cost estimates during this phase are often categorized as schematic or design development estimates. These estimates use detailed drawings and specifications to assess the costs associated with the various components of the project. The level of accuracy improves compared to the preliminary stage, but there may still be some uncertainty due to ongoing design changes.

3. Construction Document Stage

In the construction document stage, the project design is finalized, and detailed construction documents are prepared. Cost estimates at this stage are known as detailed or definitive estimates. These estimates are based on comprehensive project drawings, specifications, and material take-offs, providing a high level of accuracy.

Detailed estimates involve a thorough analysis of the quantities and costs of all project components, including labor, materials, and overheads. This stage also includes the preparation of a detailed bill of quantities, which serves as a basis for contractor pricing and tendering.

4. Tendering and Procurement Stage

During the tendering and procurement stage, cost estimates are used to evaluate contractor bids and determine the final project cost. The estimates prepared at this stage are referred to as bid estimates or tender estimates. They are based on the submitted bids from contractors and are used to assess the financial implications of the proposed construction contracts.

Bid estimates are compared against the project's budget and previous estimates to ensure that the proposed costs are reasonable and within the allocated budget. This stage may involve negotiations with contractors to refine the cost and ensure alignment with the project's financial objectives.

5. Post-Construction Stage

After the construction phase, cost estimating continues to play a role in the post-construction stage. This involves preparing final cost reports and analyzing cost performance against the initial budget. The final cost estimate includes all actual costs incurred during the construction process, including any variations, change orders, and unforeseen expenses.

The post-construction stage also includes the preparation of the final account, which provides a detailed account of the total project cost and serves as a basis for final payments and financial reconciliation. Lessons learned from cost estimating and financial management during the project are reviewed to improve future cost estimating practices and project performance.

Cost estimating at different project stages is essential for effective financial management throughout the construction project lifecycle. Each stage requires a different level of detail and accuracy, reflecting the evolving nature of the project and the increasing availability of information. By understanding and applying appropriate cost estimating techniques at each stage, project managers can ensure that the project remains on budget and achieves its financial goals. Table 8 summarizes the different approaches to cost estimating at various stages of the project lifecycle.

Table 8: Cost estimating at various stages of the project lifecycle

Project Stage	Type of Cost Estimate	Purpose	Typical Accuracy
Conceptual Stage	Preliminary Estimate	Provides a broad cost estimate for initial planning and feasibility.	±20% to ±30%
Design Stage	Detailed Estimate	Refines cost estimates based on detailed designs and specifications.	±10% to ±15%
Construction Stage	Final Estimate	Represents the final cost estimate used for budget control during construction.	±5% to ±10%

Identifying and Managing Cost Risks

Cost risk management is a crucial aspect of building cost estimating that ensures financial stability and project success. Identifying and managing cost risks involves recognizing potential uncertainties and threats that could impact the project's budget and implementing strategies to mitigate their effects. This section explores the key elements of cost risk identification and management.

1. Identifying Cost Risks

The identification of cost risks begins with a thorough analysis of the project's scope, design, and external factors. Risks can arise from various sources, including technical complexities, regulatory changes, market fluctuations, and unforeseen site conditions. The following methods are commonly employed to identify cost risks:

- **Risk Assessment Workshops**: Conducting workshops with project stakeholders, including designers, engineers, contractors, and quantity surveyors, helps identify potential risks through collaborative discussions and brainstorming sessions.

- **Historical Data Analysis**: Reviewing data from previous similar projects provides insights into common cost risks and their impact. Historical data helps in recognizing patterns and potential pitfalls that may recur.

- **Expert Judgment**: Consulting with experts in specific areas of the project, such as construction, materials, or regulations, helps identify risks based on their experience and knowledge.

- **Risk Registers**: Maintaining a risk register, a comprehensive list of identified risks along with their potential impact and likelihood, helps in tracking and managing risks throughout the project lifecycle.

2. Assessing Risk Impact and Likelihood

Once identified, each cost risk should be assessed for its potential impact on the project's budget and schedule. This assessment involves determining the likelihood of the risk occurring and the severity of its consequences. Key steps in this process include:

- **Qualitative Assessment**: Risks are categorized based on their impact and likelihood using qualitative descriptors such as high, medium, or low. This initial assessment provides a broad understanding of the risks' significance.

- **Quantitative Assessment**: For more precise risk management, quantitative methods such as Monte Carlo simulations or sensitivity analysis are used to estimate the financial impact of risks. This approach involves modeling various scenarios to predict potential cost deviations.

- **Risk Prioritization**: Risks are prioritized based on their assessed impact and likelihood. High-priority risks require more immediate attention and robust mitigation strategies, while lower-priority risks may be monitored with less intensive measures.

3. Developing and Implementing Risk Mitigation Strategies

Effective management of cost risks involves developing and implementing strategies to mitigate their impact. Mitigation strategies can be proactive or reactive and should be tailored to the specific risks identified. Common strategies include:

- **Risk Avoidance**: Modifying project plans or designs to eliminate or reduce the risk entirely. For example, choosing alternative construction methods to avoid risks associated with complex techniques.

- **Risk Transfer**: Shifting the financial responsibility of certain risks to other parties through contracts or insurance. For instance, transferring risks related to material price fluctuations to suppliers through fixed-price contracts.

- **Risk Reduction**: Implementing measures to minimize the likelihood or impact of risks. This could involve enhancing project controls, improving design accuracy, or adopting advanced technologies to reduce uncertainty.

- **Risk Acceptance**: Acknowledging the risk and preparing to manage its consequences if it occurs. This approach is often used for risks with low impact or those where mitigation measures are not cost-effective.

4. Monitoring and Reviewing Risks

Continuous monitoring and review of cost risks are essential to ensure that mitigation strategies remain effective and relevant. Regular risk reviews involve:

- **Periodic Risk Assessments**: Updating the risk register and reassessing risks at various stages of the project to reflect any changes in project conditions or external factors.

- **Risk Reporting**: Providing regular reports on risk status to project stakeholders, including information on risk occurrences, mitigation efforts, and their impact on the project budget.

- **Lessons Learned**: Analyzing and documenting lessons learned from risk management efforts to improve future risk management practices and project performance.

Identifying and managing cost risks is integral to maintaining financial control and ensuring project success. By systematically identifying potential risks, assessing their impact and likelihood, developing effective mitigation strategies, and continuously monitoring risks, project managers can safeguard against cost overruns and achieve the project's financial objectives. Table 9 identifies common cost risks in construction projects and provides mitigation strategies to address each risk.

Table 9: Common cost risks in construction projects

Cost Risk	Description	Mitigation Strategy
Scope Changes	Unplanned alterations to the project scope	Implement robust change management processes; Ensure clear scope definition.
Price Fluctuations	Variability in material or labor costs	Utilize fixed-price contracts where feasible; Monitor market trends for early warning.

The Importance of Cost Control Throughout the Project Lifecycle

Cost control is a fundamental aspect of project management that ensures a construction project remains within its budgetary constraints. It involves the systematic monitoring and regulation of project costs to prevent deviations from the planned financial parameters. Effective cost control is crucial throughout the entire project lifecycle, from initial planning to project completion. This section delves into the significance of cost control and its various components.

1. Role of Cost Control in Project Success

Cost control is pivotal in achieving project success by ensuring that financial resources are utilized efficiently and effectively. It contributes to several key aspects of project management:

- **Budget Adherence**: Cost control helps maintain the project's financial health by preventing budget overruns. By monitoring expenditures and comparing them to the budgeted amounts, project managers can take corrective actions to address any deviations.

- **Financial Forecasting**: Ongoing cost control provides valuable data for forecasting future financial requirements. Accurate forecasts enable project managers to allocate resources appropriately and plan for potential cost adjustments.

- **Resource Optimization**: Effective cost control ensures that resources are used optimally. It involves analyzing resource utilization, identifying cost-saving opportunities, and avoiding wastage, thereby enhancing overall project efficiency.

- **Risk Management**: By closely monitoring costs, project managers can identify potential financial risks early and implement mitigation strategies. This proactive approach helps in managing uncertainties and reducing the impact of unexpected cost changes.

2. Cost Control Strategies

Implementing robust cost control strategies is essential for managing project expenses effectively. These strategies include:

- **Cost Planning**: Establishing a detailed cost plan at the project's outset is crucial. This plan includes cost estimates for various project components, a budget allocation, and financial controls. A well-defined cost plan serves as a baseline for tracking actual expenditures and identifying variances.

- **Budget Monitoring**: Regularly monitoring actual costs against the budgeted amounts is vital for effective cost control. This involves tracking expenditures, reviewing financial reports, and comparing them to the planned budget. Variance analysis helps in identifying discrepancies and taking corrective actions.

- **Cost Reporting**: Regular cost reports provide an overview of the project's financial status. These reports should include detailed breakdowns of costs, budget variances, and explanations for any deviations. Transparent reporting ensures that stakeholders are informed and can make informed decisions.

- **Change Management**: Managing changes to the project scope, design, or schedule is crucial for cost control. Any changes should be evaluated for their impact on the budget, and adjustments should be made accordingly. Change orders should be documented and approved to avoid unauthorized cost increases.

- **Cost Control Tools**: Utilizing cost control tools and software can enhance the accuracy and efficiency of financial management. These tools offer features for tracking expenses, generating reports, and analyzing financial data, aiding in better decision-making.

3. Cost Control Throughout the Project Lifecycle

Cost control is not a one-time activity but an ongoing process that spans the entire project lifecycle. Key phases where cost control is applied include:

- **Pre-Construction Phase**: During the planning and design stages, cost control involves developing accurate cost estimates, establishing budgets, and planning for potential financial risks. This phase sets the foundation for effective cost management throughout the project.

- **Construction Phase**: In this phase, cost control focuses on monitoring actual expenditures, managing changes, and addressing any cost-related issues that arise. Regular cost reviews and adjustments help ensure that the project stays within budget.

- **Post-Construction Phase**: Even after construction is complete, cost control continues through the final accounting and settlement of project costs. This phase involves reconciling the final costs with the budget, settling outstanding payments, and analyzing financial performance.

4. Benefits of Effective Cost Control

Effective cost control offers several benefits, including:

- **Enhanced Financial Performance**: By keeping costs within budget, projects achieve better financial outcomes, contributing to overall profitability and success.

- **Improved Decision-Making**: Accurate cost data and reporting enable informed decision-making, allowing project managers to address issues proactively and make strategic adjustments.

- **Increased Stakeholder Confidence**: Transparent cost control practices build trust and confidence among stakeholders, including clients, investors, and project teams.

- **Successful Project Delivery**: Adhering to budget constraints and managing costs effectively contributes to the successful delivery of the project, meeting client expectations and project objectives.

Cost control is a critical component of project management that ensures financial discipline throughout the project lifecycle. By implementing effective cost control strategies, monitoring expenditures, and addressing cost-related issues promptly, project managers can achieve budget adherence, optimize resource utilization, and enhance overall project success. The continuous application of cost control practices contributes to the financial health of the project and supports its successful completion.

CHAPTER 5:
PREPARING BILLS
OF QUANTITIES

The Purpose and Structure of BOQs

Bills of Quantities are essential documents in the construction industry, serving as a detailed and structured summary of the work to be carried out on a project. Their primary purpose is to provide a comprehensive and clear representation of the quantities of materials, labor, and other resources required for the successful completion of a construction project. By breaking down the work into manageable and quantifiable sections, BOQs facilitate accurate cost estimation, effective project planning, and precise tendering processes.

The structure of a Bill of Quantities is designed to ensure that all aspects of the project are covered systematically. It typically includes several key sections:

- **Preliminaries**: This section outlines the general requirements of the project, including administrative and regulatory aspects. It often details the project's scope, specifications, and conditions that apply throughout the construction process.

- **Quantities**: This is the core section of the BOQ, where the quantities of each item of work are listed. Each item is described in detail, including its location, dimensions, and any relevant specifications. Quantities are usually measured in standard units, such as square meters for flooring or cubic meters for concrete.

- **Descriptions**: Each item in the BOQ is accompanied by a description that specifies the nature of the work required. This helps to ensure that all parties involved have a clear understanding of what is expected and reduces the potential for disputes or misunderstandings.

- **Rates and Costs**: While not always included in the initial BOQ, rates and costs may be added later as part of the tendering process. This section provides the unit rates for each item of work and calculates the total cost based on the quantities specified.

- **Summary**: The summary section aggregates the quantities and costs from the individual items, providing an overall view of the project's scope and budget. This helps in comparing different tenders and assessing the overall project cost.

The structured approach of a BOQ ensures that it serves as a comprehensive tool for communication between all stakeholders, including clients, contractors, and quantity surveyors. By providing a clear and detailed breakdown of the work, it enables more accurate pricing, effective project management, and better control over the project's budget and schedule.

Creating Clear and Accurate BOQs

The creation of clear and accurate Bills of Quantities is crucial for the successful execution of construction projects. A well-prepared BOQ serves as a reliable reference document that facilitates precise cost estimation, effective project planning, and transparent communication between parties. To achieve clarity and accuracy in a BOQ, several key principles and practices must be adhered to:

1. Detailed Descriptions: Each item listed in the BOQ should be described in detail, including precise measurements, materials, and construction methods. Clear descriptions help avoid ambiguities and ensure that all parties understand the scope of work. For instance, instead of simply stating "brickwork," the description should specify "brickwork with 1:6 cement mortar, including all necessary lintels and ties."

2. Consistent Measurement Units: Accurate measurement is fundamental to a BOQ's clarity. Consistent use of measurement units, such as square meters for plastering or cubic meters for concrete, helps maintain uniformity throughout the document. It is also important to use standard units and conventions, as specified by industry guidelines, to facilitate comparison and verification.

3. Comprehensive Scope: The BOQ should cover all aspects of the project, including preliminaries, detailed work items, and any additional requirements. Omissions can lead to discrepancies between the BOQ and actual work, resulting in disputes or unexpected costs. A comprehensive scope ensures that all aspects of the project are accounted for, reducing the risk of incomplete or inaccurate pricing.

4. Clear Formatting: The layout and formatting of the BOQ should be clear and organized. Each section should be logically structured, with headings, subheadings, and item numbers that make it easy to navigate. Consistent formatting helps users quickly locate specific items and understand the overall structure of the BOQ.

5. Accuracy in Quantification: Precise measurement and calculation are essential for accuracy. Errors in quantification can lead to significant discrepancies in cost estimation and project budgeting. Therefore, careful measurement and verification of quantities are necessary before finalizing the BOQ.

6. Regular Updates: As the project progresses, changes to the scope of work or design may occur. It is important to regularly update the BOQ to reflect any modifications. This ensures that the document remains accurate and relevant throughout the project's lifecycle.

7. Collaboration with Stakeholders: Effective communication with project stakeholders, including architects, engineers, and contractors, is vital in creating an accurate BOQ. Collaboration helps ensure that all requirements are captured and that any potential issues are addressed before finalizing the document.

By adhering to these principles, quantity surveyors can create BOQs that serve as reliable tools for cost estimation and project management. Clear and accurate BOQs contribute to better project outcomes, reducing the likelihood of disputes and ensuring that the project stays within budget and on schedule.

Using Standard BOQ Templates

The use of standard Bills of Quantities templates is an essential practice in the construction industry. These templates provide a structured framework that helps ensure consistency, accuracy, and completeness in the preparation of BOQs. Employing standard templates can significantly streamline the process of BOQ preparation, making it more efficient and less prone to errors.

1. Benefits of Using Standard Templates

Standard BOQ templates offer several advantages. They help maintain uniformity across different projects, ensuring that all relevant information is included and presented consistently. This consistency facilitates easier comparison of BOQs from different projects and contractors, which can be especially useful in competitive bidding processes. Additionally, standard templates reduce the likelihood of omissions and inaccuracies by providing a pre-defined structure that covers all necessary aspects of the BOQ.

2. Components of a Standard BOQ Template

A typical standard BOQ template includes several key components:

- **Header Information**: This section includes project details, such as the project name, location, client, and the name of the quantity surveyor. It may also include references to the relevant contract documents and standards.

- **Preliminaries**: This section outlines the preliminary requirements for the project, such as site preparation, insurance, and any specific conditions that must be met before the main work can commence.

- **Work Sections**: These are divided into various categories based on the types of work involved, such as excavation, concrete work, and finishes. Each section typically includes a list of items with detailed descriptions, quantities, and units of measurement.

- **Item Descriptions**: For each work section, the template provides space for detailed item descriptions, including the scope of work, materials, and construction methods. This ensures that each item is clearly defined and accurately quantified.

- **Quantities and Units**: The template includes columns for recording quantities and units of measurement. Accurate quantification is crucial for effective cost estimation and budgeting.

- **Pricing and Totals**: This section allows for the insertion of unit rates and total costs for each item. It also includes a summary of total costs for each work section and the overall project.

- **Notes and Clarifications**: There is often a section for additional notes or clarifications that provide further details or specific instructions related to the BOQ items.

In summary, key elements of a standard BoQ template include:

- **Project identification**: Project name, client, location, and contract number.
- **Preparation date**: Date of BoQ preparation.
- **Measurement method**: Specified measurement standards used.
- BoQ structure: Clear divisions and sections for different project elements.
- **Item description**: Detailed and unambiguous description of each item.
- **Units of measurement**: Consistent use of standard units.
- **Quantities**: Accurate measurement of quantities for each item.
- **Rates and prices**: Columns for unit rates and total costs.
- **Totals and summaries**: Summary of total quantities and costs for each section and overall project.

By utilizing a standard BoQ template, quantity surveyors can streamline the preparation process, reduce errors, and enhance the clarity of the document. Table 10 presents a detailed overview of the typical structure of a Bill of Quantities (BOQ). It outlines the various sections and components that are commonly included

Table 10: BOQ Structure Overview

Item Description	Detailed description of the work or materials required
Quantity	Amount or volume of work to be performed or materials needed
Unit Rate	Cost per unit of measurement (e.g., per square meter)
Total Cost	Calculated cost based on quantity and unit rate
Subtotal	Sum of costs for specific sections or categories
Grand Total	Total cost for the entire BoQ

3. Adapting Templates to Specific Projects

While standard BOQ templates provide a useful starting point, they may need to be adapted to suit the specific requirements of individual projects. Adjustments might be necessary to accommodate unique project specifications, local regulations, or particular client needs. Quantity surveyors should review and customize the template as needed to ensure it accurately reflects the project's scope and requirements.

4. Ensuring Compliance with Standards

It is important to ensure that the standard BOQ templates comply with relevant industry standards and guidelines. Adhering to established standards helps maintain consistency and quality in BOQ preparation. Quantity surveyors should be familiar with the applicable standards and incorporate them into the template as required.

5. Benefits of Template Use in Practice

The adoption of standard BOQ templates streamlines the BOQ preparation process, saving time and reducing the risk of errors. It enhances clarity and transparency, facilitating effective communication between project stakeholders. By providing a structured approach to BOQ preparation, templates contribute to more accurate cost estimation and better project management. Using standard BOQ templates is a best practice in the field of quantity surveying. It ensures consistency, accuracy, and completeness in the preparation of BOQs, ultimately contributing to successful project outcomes.

Common Errors in Preparing BOQs and How to Avoid Them

Preparing Bills of Quantities (BOQs) is a critical task in construction projects that requires meticulous attention to detail. Despite the structured nature of the process, several common errors can arise during the preparation of BOQs, leading to potential issues in cost estimation, project budgeting, and overall project execution. Understanding these errors and implementing strategies to avoid them is essential for ensuring accuracy and reliability in BOQ preparation.

1. Omissions and Inaccuracies

One of the most prevalent errors in BOQ preparation is the omission of items or inaccuracies in the descriptions and quantities. Omissions can occur due to oversight or incomplete information, while inaccuracies might result from incorrect measurements or misunderstandings of the project requirements. To mitigate these issues, it is crucial to thoroughly review project drawings and specifications, cross-check quantities with site measurements, and ensure that all relevant items are included in the BOQ.

2. Inconsistent Units of Measurement

Inconsistencies in units of measurement can lead to confusion and errors in cost estimation. For example, using different units (such as square meters versus square feet) within the same BOQ can result in incorrect quantity calculations and pricing discrepancies. To avoid this error, it is essential to use standardized units of measurement throughout the BOQ and ensure that all measurements are consistently applied. Employing standard measurement practices and referencing relevant guidelines can help maintain uniformity.

3. Incorrect Pricing

Errors in pricing can significantly impact the accuracy of the BOQ and the overall project budget. Pricing errors can arise from miscalculations, incorrect unit rates, or failure to account for all associated costs. To prevent pricing errors, quantity surveyors should carefully verify unit rates, consider all cost factors (including labor, materials, and overheads), and apply appropriate mark-ups. It is also advisable to review and update pricing data regularly to reflect current market conditions.

4. Lack of Clarity in Item Descriptions

Ambiguous or unclear item descriptions can lead to misunderstandings between the quantity surveyor and other project stakeholders. Vague descriptions may result in variations or disputes during the project. To ensure clarity, item descriptions should be detailed and specific, providing comprehensive information about the scope of work, materials, and construction methods. Clear and precise language helps prevent misinterpretations and facilitates accurate cost estimation.

5. Failure to Update BOQs

BOQs should be updated regularly to reflect changes in the

project scope, design modifications, or variations. Failure to update the BOQ can lead to discrepancies between the BOQ and actual project requirements, resulting in cost overruns and delays. Implementing a systematic approach to tracking and documenting changes, and regularly reviewing the BOQ against project progress, helps maintain accuracy and alignment with the current project status.

6. Inadequate Review and Quality Control

The absence of thorough review and quality control processes can lead to undetected errors in the BOQ. Inadequate review may result in overlooked mistakes or inconsistencies. To avoid this, it is essential to implement a robust review process that involves multiple checks and verification steps. Peer reviews, cross-referencing with project documents, and using checklists can help identify and rectify errors before finalizing the BOQ.

7. Ignoring Local Regulations and Standards

Different regions and jurisdictions may have specific regulations and standards that impact BOQ preparation. Ignoring these requirements can result in non-compliance and potential legal issues. Quantity surveyors should be well-versed in local regulations and ensure that the BOQ adheres to relevant standards and guidelines. Staying informed about regulatory changes and incorporating them into the BOQ preparation process is crucial for compliance.

8. Inadequate Documentation

Proper documentation is essential for maintaining the accuracy and transparency of the BOQ. Inadequate documentation may result in missing or incomplete information, leading to misunderstandings or disputes. Comprehensive documentation should include detailed item descriptions, supporting calculations, and any relevant references or notes. Maintaining organized and thorough documentation supports the credibility and reliability of the BOQ.

Avoiding common errors in BOQ preparation requires a proactive approach to accuracy, consistency, and quality control. By addressing these issues and implementing best practices, quantity surveyors can enhance the reliability of the BOQ and contribute to successful project outcomes. Table 11 summarizes common errors encountered during the preparation of Bills of Quantities and provides practical strategies for avoiding these issues.

Table 11: Common errors in the preparation of Bills of Quantities

Error Type	Description	Avoidance Strategy
Omissions and Inaccuracies	Missing or incorrect quantities	Thorough review of drawings and specifications; cross-check quantities
Inconsistent Units of Measurement	Use of different units within the BoQ	Standardize units of measurement; ensure consistency throughout
Incorrect Pricing	Errors in unit rates or pricing calculations	Verify unit rates; consider all cost factors; update pricing data
Lack of Clarity in Item Descriptions	Ambiguous or incomplete descriptions	Provide detailed and specific item descriptions
Failure to Update BoQs	Not reflecting changes in project scope or design modifications	Regularly update the BoQ to reflect changes and track progress
Inadequate Review and Quality Control	Insufficient checks leading to undetected errors	Implement thorough review processes and quality control measures
Ignoring Local Regulations and Standards	Non-compliance with regional regulations	Adhere to local regulations and standards; stay informed about changes
Inadequate Documentation	Missing or incomplete documentation	Ensure comprehensive and organized documentation is maintained

PART 3: QUANTITY SURVEYING THROUGHOUT THE PROJECT LIFECYCLE

CHAPTER 6: QUANTITY SURVEYING IN PRE-CONSTRUCTION

Cost Planning and Feasibility Studies

In the pre-construction phase, cost planning and feasibility studies are essential for establishing a project's financial viability and laying the groundwork for effective budget management throughout the project's lifecycle. This stage involves a comprehensive analysis of potential costs, resource allocation, and economic factors that may impact the project.

Cost planning begins with the development of a detailed budget, which serves as a financial roadmap for the project. This budget is based on preliminary designs and includes estimates for all aspects of the project, such as materials, labor, equipment, and overhead costs. The accuracy of these estimates is crucial, as they form the basis for financial decision-making and risk management. Quantity surveyors play a vital role in this process, using their expertise to provide detailed cost estimates and ensure that all potential expenses are accounted for.

Feasibility studies complement cost planning by evaluating

the project's overall viability. These studies assess various factors, including site conditions, regulatory requirements, environmental impact, and market demand. The goal is to determine whether the project is technically, economically, and legally feasible. This involves a thorough analysis of potential risks and benefits, as well as an evaluation of alternative solutions or approaches. The findings from feasibility studies inform key decisions, such as project design, funding strategies, and timelines.

In addition to these analyses, cost planning and feasibility studies often involve the development of cost control measures. These measures are designed to monitor and manage costs throughout the project, ensuring that expenditures remain within the approved budget. This includes establishing cost baselines, identifying potential cost-saving opportunities, and implementing cost-tracking systems. Effective cost control is essential for maintaining financial discipline and preventing budget overruns.

One of the challenges in this phase is the uncertainty inherent in early-stage project planning. The initial estimates are often based on limited information, making it challenging to predict all potential costs accurately. To mitigate this uncertainty, quantity surveyors use various estimation techniques, such as cost modeling and benchmarking, to provide a range of cost scenarios. This approach helps stakeholders understand the potential financial implications under different conditions and make informed decisions.

Cost planning and feasibility studies ensure that all potential costs are considered, the project's viability is thoroughly evaluated, and cost control measures are in place. By providing accurate estimates and analysis, quantity surveyors enable stakeholders to make informed decisions, manage risks, and ensure the project's financial success.

Tendering and Procurement Processes

The tendering and procurement processes are fundamental stages in the pre-construction phase, where contractors and suppliers are selected for the project. These processes ensure that the project secures the necessary resources and expertise at competitive prices, while also maintaining quality and compliance standards.

Tendering Process

The tendering process begins with the preparation of tender documents, which are detailed specifications that outline the scope of work, technical requirements, project timelines, and contractual terms. These documents also include drawings, bills of quantities, and any other relevant information needed by potential bidders to prepare their proposals. The accuracy and clarity of tender documents are crucial, as they serve as the basis for bids and subsequent contractual agreements.

Once the tender documents are ready, the project is advertised to invite bids from interested contractors and suppliers. This can be done through public advertisements, direct invitations to pre-qualified bidders, or through online procurement platforms. The selection of the tendering method—open, selective, or negotiated—depends on the project's nature, complexity, and specific requirements.

Bidders submit their proposals, which typically include cost estimates, work schedules, technical capabilities, and compliance with legal and safety standards. The bids are evaluated based on various criteria, including price, quality, experience, and the ability to meet project timelines. The evaluation process is designed to ensure transparency, fairness, and competitiveness, with the ultimate goal of selecting the best value offer.

The tendering process typically involves the following steps:

- **Preparation of tender documents**: Developing comprehensive tender documents, including the bill of quantities, contract conditions, drawings, and specifications.
- **Tender advertising**: Disseminating the tender documents to potential bidders through appropriate channels.
- **Pre-qualification of bidders**: Evaluating the capabilities and financial standing of interested contractors.
- **Tender submission**: Receiving and opening bids from qualified contractors.
- **Tender evaluation**: Analyzing and comparing bids based on price, quality, and other criteria.
- **Contract award**: Selecting the successful bidder and issuing a contract.

Procurement Process

Procurement involves the actual acquisition of goods, services, and works necessary for the project. This process encompasses the selection of suppliers, negotiation of contracts, and the management of relationships with vendors. Procurement strategies vary depending on the project's needs, the market conditions, and the risk profile.

There are several procurement methods, including traditional procurement, design and build, management contracting, and construction management. Each method has its advantages and drawbacks, and the choice depends on factors such as project complexity, time constraints, and the degree of control required by the client. Traditional procurement, for example, separates the design and construction phases, allowing for detailed design development before contractor selection. In contrast, design and build integrate these phases, potentially speeding up the project but requiring a higher level of coordination and trust.

During the procurement process, it is essential to establish clear contractual terms, including payment schedules, quality standards, and dispute resolution mechanisms. Effective contract management is crucial to ensure that all parties meet their obligations and that any issues are addressed promptly. Quantity surveyors often play a key role in contract administration, monitoring compliance, and managing changes to the contract scope.

Additionally, procurement involves the assessment of suppliers' financial stability, capacity, and past performance. This assessment helps mitigate risks related to delays, non-performance, or quality issues. Maintaining strong relationships with suppliers is also vital, as it can lead to better pricing, more reliable delivery schedules, and improved overall project outcomes.

Tendering and procurement require careful planning, transparent evaluation, and effective management to ensure that the project is executed on time, within budget, and to the required standards. Quantity surveyors play a vital role in these processes, from preparing tender documents to overseeing contract administration, ensuring that the project's financial and operational goals are achieved.

Analyzing Contractor Bids

Analyzing contractor bids is a crucial phase in the pre-construction process, involving the detailed examination and comparison of proposals submitted by contractors. This process aims to identify the most suitable contractor based on a combination of factors including cost, quality, experience, and compliance with project requirements. The analysis ensures that the selected contractor can deliver the project effectively and efficiently, aligning with the client's objectives and expectations.

Initial Screening and Compliance Check

The first step in analyzing contractor bids is to perform an initial screening to ensure compliance with the basic requirements outlined in the tender documents. This involves checking whether the bids include all necessary documents and information, such as cost estimates, work schedules, and qualifications. Bids that do not meet these requirements may be disqualified or set aside for further clarification. This stage helps to filter out non-compliant or incomplete bids, ensuring that only those meeting the minimum standards are considered.

Detailed Cost Analysis

A detailed cost analysis involves scrutinizing the cost breakdown provided by each contractor. This analysis goes beyond the total bid price to assess the accuracy and reasonableness of individual cost components, such as labor, materials, equipment, and overheads. The purpose is to identify any anomalies, such as unusually low or high prices for certain items, which could indicate potential issues like underestimation, misunderstanding of the scope, or attempts at strategic bidding.

Quantity surveyors play a key role in this phase, leveraging their expertise to evaluate whether the proposed costs are realistic and aligned with market rates. They may also use benchmarking techniques, comparing the bids against historical data or similar projects to gauge the competitiveness of the prices.

Technical and Qualitative Assessment

Beyond cost, the technical and qualitative aspects of the bids are equally important. This assessment examines the contractor's proposed methodology, work plan, and schedule to determine their feasibility and alignment with the project's requirements. It also involves evaluating the contractor's experience, past performance, and technical expertise.

A contractor's experience can be assessed based on their track record in delivering similar projects, including their ability to manage complexity, adhere to timelines, and maintain quality standards. References from previous clients, as well as performance metrics from completed projects, can provide valuable insights into a contractor's reliability and competence.

Risk Assessment and Financial Stability

Analyzing contractor bids also involves assessing potential risks associated with each bid. This includes evaluating the contractor's financial stability, insurance coverage, and safety record. Financial stability is particularly important, as contractors with weak financial positions may struggle to mobilize resources, manage cash flow, or withstand unforeseen challenges, leading to delays or quality issues.

Contractors' safety records and compliance with health and safety regulations are also critical considerations. A poor safety record may indicate a higher risk of accidents, potentially leading to project delays, increased costs, or legal liabilities.

Value for Money and Best Value Selection

The final stage of analyzing contractor bids is to consider the overall value for money. This involves balancing cost with quality, experience, and risk factors to identify the bid that offers the best value, rather than simply the lowest price. The selection process should aim to ensure that the chosen contractor can deliver the project to the required standards, on time, and within budget, while also managing risks effectively.

In some cases, a best value selection approach may be adopted, where a scoring system is used to evaluate the bids against predefined criteria. This approach allows for a more structured and transparent evaluation, ensuring that all relevant factors are considered in the decision-making process.

Analyzing contractor bids is a complex and multi-faceted process that requires a thorough understanding of both financial and technical aspects. It involves balancing cost considerations with qualitative factors such as experience, capability, and risk management. By conducting a comprehensive analysis, clients and quantity surveyors can select a contractor that not only offers competitive pricing but also demonstrates the capability and reliability to deliver the project successfully. This careful and methodical approach is essential for mitigating risks and ensuring the project's success from the outset.

Key elements of bid analysis include:

- **Bid completeness**: Ensuring that all required information and documentation are included in the bid.
- **Price analysis**: Comparing bid prices to the project budget and identifying any significant deviations.
- **Commercial evaluation**: Assessing the commercial terms and conditions of the bid, including payment terms, warranties, and liabilities.
- **Technical evaluation**: Evaluating the contractor's proposed methods, resources, and qualifications.
- **Risk assessment**: Identifying potential risks associated with each bid.

Table 12 highlights the key factors to consider during the tendering and procurement processes.

Table 12: Tendering and procurement factors

Factor	Description
Scope Definition	Clearly defines project requirements and specifications, ensuring all bidders understand the project scope and eliminates ambiguity in pricing.
Procurement Method Selection	Involves choosing the most appropriate procurement method (e.g., traditional, design-build, construction management) based on project complexity, risk profile, and client objectives.
Budget and Financial Assessment	Assesses project budget constraints and financial feasibility, ensuring the availability of funds throughout the project lifecycle to avoid cost overruns and delays.
Risk Assessment and Mitigation	Identifies potential risks associated with the project (e.g., material price fluctuations, labor shortages) and outlines strategies to mitigate these

	risks during the procurement phase.
Legal and Regulatory Compliance	Ensures all tender documents and processes comply with relevant laws, regulations, and industry standards to minimize legal challenges and project delays.
Timeline and Schedule Management	Develops a realistic project timeline, including key milestones and deadlines, for the procurement process to ensure timely completion and avoid disruptions.
Stakeholder Communication	Facilitates transparent communication with all stakeholders involved in the procurement process, including clients, contractors, and consultants. This ensures alignment of expectations and requirements, leading to a smoother procurement process.

Strategies for Effective Tender Evaluation and Contractor Selection

Selecting the right contractor is a critical decision in the pre-construction phase, influencing the overall success of a project. An effective tender evaluation process ensures that the selected contractor is not only capable of delivering the project within budget and on time but also adheres to the desired quality and safety standards. This section outlines key strategies for conducting a thorough and fair evaluation of tenders and making an informed contractor selection.

Developing a Clear Evaluation Framework

The foundation of effective tender evaluation lies in developing a clear and comprehensive evaluation framework. This framework should outline the criteria against which all bids will be assessed, ensuring consistency and transparency in the evaluation process. Key criteria typically include cost, technical capability, experience, safety record, and financial stability. Additionally, criteria specific to the project's needs, such as sustainability practices or innovation in construction methods, may also be included.

Each criterion should be clearly defined, with specific metrics or benchmarks established for evaluation. For example, cost criteria might include both initial bid price and life-cycle costs, while technical capability might be assessed based on the proposed construction methodology and work plan. By setting these benchmarks in advance, evaluators can maintain objectivity and avoid biases in the assessment process.

Weighting and Scoring System

Implementing a weighting and scoring system is an effective strategy for quantifying the evaluation process. This system assigns relative importance to each evaluation criterion through weighting, reflecting the project's priorities and objectives. For instance, in a project where quality is paramount, the technical capability might be given a higher weight compared to cost.

Each bid is then scored against the criteria, with the scores multiplied by the respective weights to derive a weighted score. The total weighted score for each bid provides a quantitative measure of its overall value. This approach allows for a structured comparison of bids, facilitating the identification of the contractor that offers the best overall value, rather than simply the lowest price.

Pre-Qualification and Shortlisting

Pre-qualification is a preliminary step that can streamline the tender evaluation process. This involves assessing contractors' general capabilities and suitability for the project before they are invited to submit detailed bids. Pre-qualification criteria may include factors such as experience in similar projects, financial stability, availability of key resources, and compliance with regulatory standards.

Once pre-qualification is complete, a shortlist of qualified contractors is created. This shortlisting process ensures that only those contractors who meet the basic requirements and have demonstrated the necessary expertise are considered, reducing the time and effort required for detailed evaluation.

Conducting Detailed Bid Analysis

A detailed analysis of the shortlisted bids is essential for a thorough evaluation. This analysis should encompass both quantitative and qualitative aspects. Quantitative analysis involves examining the cost breakdown, identifying any anomalies or potential savings, and assessing the overall financial viability of the bid. Qualitative analysis focuses on the contractor's technical proposal, work plan, and project management approach, evaluating their ability to meet the project's technical and logistical challenges.

During this stage, it is important to consider the alignment between the contractor's proposal and the project's goals. This includes assessing the contractor's understanding of the project requirements, proposed innovations or improvements, and strategies for risk management and problem-solving.

Interviews and Negotiations

Interviews and negotiations can be valuable components of the evaluation process, providing an opportunity to clarify any ambiguities in the bids and assess the contractors' communication and problem-solving skills. Interviews allow evaluators to pose specific questions about the bid, such as how the contractor plans to address potential risks or manage key project milestones.

Negotiations may also be used to refine the terms of the contract, address any concerns, and ensure mutual understanding between the client and contractor. This process can help in securing more favorable terms, such as better pricing, improved timelines, or enhanced quality guarantees.

Final Selection and Due Diligence

The final selection should be based on the comprehensive evaluation of all factors, including cost, quality, capability, and risk. The contractor with the highest weighted score or the best overall evaluation should be selected, provided they meet all necessary requirements and have demonstrated the ability to deliver the project successfully.

Due diligence is a critical final step, involving the verification of the contractor's credentials, references, and previous project performance. This ensures that the selected contractor has a proven track record and can be relied upon to fulfill their obligations under the contract.

By employing a structured evaluation framework, a scoring system, and thorough due diligence, project managers and quantity surveyors can make informed decisions that align with project objectives. This careful selection process not only ensures the best value for money but also contributes to the overall quality, efficiency, and success of the project. Table 13 summarizes the key strategies for effective tender evaluation and contractor selection.

Table 13: Tender evaluation and contractor selection

Strategy	Description
Clear Evaluation Framework	Establishes consistent criteria for assessing bids, including cost, technical capability, experience, safety, and financial stability.
Weighting and Scoring System	Assigns relative importance to each evaluation criterion, allowing for a quantitative comparison of bids based on a weighted score.
Pre-Qualification and Shortlisting	Filters out unsuitable contractors at an early stage, focusing detailed evaluation on a shortlist of qualified bidders.
Detailed Bid Analysis	Involves both quantitative (cost breakdown, financial viability) and qualitative (technical proposal, work plan) analysis of bids.
Interviews and Negotiations	Provides an opportunity to clarify bid details, assess communication skills, negotiate better terms, and enhance mutual understanding between client and contractor.
Final Selection and Due Diligence	Based on the comprehensive evaluation, verifies the credentials and previous performance of the selected contractor, ensuring their reliability and capability.

CHAPTER 7: QUANTITY SURVEYING DURING CONSTRUCTION

*Variation Orders and
Change Management*

During the construction phase, unforeseen circumstances or additional client requests often necessitate changes to the original scope of work. These modifications, known as variation orders, require careful management to ensure they are executed smoothly without negatively impacting the project's cost, schedule, or quality. Variation orders can arise from design changes, unexpected site conditions, regulatory requirements, or client-initiated alterations.

Effective change management begins with a clear and thorough documentation process, starting from the initial request for a change to its final approval. Quantity surveyors play a crucial role in evaluating the financial implications of these variations. This involves assessing the additional costs associated with labor, materials, and time. Accurate cost estimation for variations is vital to prevent budget overruns and ensure that the project remains financially viable.

Moreover, managing variation orders requires clear communication and coordination among all stakeholders, including the client, contractor, architect, and any relevant regulatory bodies. Quantity surveyors must ensure that all parties are fully informed of the changes, their implications, and the agreed-upon adjustments. This transparency helps in minimizing disputes and maintaining a collaborative project environment.

In addition to financial considerations, quantity surveyors must also evaluate the potential impact of variations on the project's timeline. Delays resulting from variations can lead to additional costs and affect the overall project schedule. Therefore, it is essential to include a time impact analysis in the variation assessment process. This analysis helps in understanding how the variation will affect the project's critical path and in determining appropriate adjustments to the schedule.

Managing variation orders and changes during construction is a complex process that requires meticulous planning, clear communication, and a comprehensive understanding of both cost and time implications. Quantity surveyors must adopt a proactive approach, anticipating potential changes and preparing to manage them efficiently. By doing so, they play a vital role in ensuring the successful delivery of construction projects, both on time and within budget. Table 14 presents a structured overview of the process involved in managing variation orders and changes during construction

Table 14: Managing variation orders and changes

Step	Description	Responsible Parties	Documents/Records
Initiation	Identification of the need for a	Contractor, Project Manager	Change Request Form

	change		
Assessment	Evaluation of the impact of the change on scope and cost	Quantity Surveyor, Project Manager	Impact Analysis Report
Approval	Formal approval or rejection of the change	Client, Project Manager	Approved Change Order
Implementation	Execution of the approved change	Contractor	Updated Construction Documents
Documentation	Recording and updating of records	Quantity Surveyor, Project Manager	Change Log, Updated BoQs

Measurement and Valuation of Construction Work

Measurement and valuation of construction work are critical aspects of quantity surveying during the construction phase. These processes ensure that the work performed is accurately assessed, recorded, and paid for, reflecting the true scope and quality of the construction activities. Accurate measurement and valuation not only help in maintaining the financial integrity of the project but also facilitate transparent and fair dealings between the contractor and the client.

Measurement of Construction Work

Measurement involves quantifying the various elements of the construction work carried out on-site. This task requires a thorough understanding of construction drawings, specifications, and the methods of measurement specified in the contract. The Standard Method of Measurement (SMM) or the New Rules of Measurement (NRM) are commonly used guidelines that provide detailed rules for measuring various aspects of construction work, such as earthworks, concrete, masonry, and finishes.

Quantity surveyors must ensure that measurements are taken accurately and systematically. This includes regular site visits to verify the work completed and to ensure that it conforms to the contract specifications. The measurements are then recorded in a format known as the Measurement Sheet, which serves as a detailed account of all work executed. This documentation is essential for preparing interim valuations, final accounts, and for resolving any disputes that may arise regarding the scope of work.

Valuation of Construction Work

Valuation, on the other hand, involves determining the monetary value of the work completed. This process is closely tied to the measurement process, as it relies on the accurate quantification of work to establish its value. Valuation is typically conducted at regular intervals throughout the construction phase, usually on a monthly basis, to facilitate progress payments to the contractor. These progress payments are based on the work completed to date and provide the contractor with the necessary cash flow to continue the project.

The valuation process includes assessing the work done, calculating the value of completed work, and deducting any retention sums, previous payments, or other deductions as specified in the contract. Quantity surveyors must ensure that valuations are fair and reflective of the actual work done, taking into account the quality and compliance with the contract requirements.

Interim and Final Valuations

Interim valuations are prepared periodically and provide a snapshot of the work completed and its corresponding value at a particular point in time. These valuations are crucial for maintaining the financial health of the project and ensuring that contractors receive timely payments for the work performed.

The final valuation, also known as the final account, is prepared after the project and represents the total value of all work carried out, including any variations and adjustments. This final account is a comprehensive statement that reconciles all payments made during the project, ensuring that the contractor is compensated for the full value of the work done.

The measurement and valuation of construction work are integral to the role of a quantity surveyor. These processes ensure that the project remains on budget, that the contractor is fairly compensated, and that the client receives value for money. By meticulously measuring and valuing the work, quantity surveyors contribute to the transparent and efficient financial management of construction projects.

Progress Claims and Payment Applications

Progress claims and payment applications are essential components of the construction project's financial management, providing a structured mechanism for contractors to receive payments for work completed at various stages of the project. These processes ensure that contractors are adequately compensated for the work carried out, enabling them to maintain cash flow and meet their financial obligations, such as paying subcontractors, suppliers, and labor.

Progress Claims

A progress claim, also known as a progress payment application, is a formal request submitted by the contractor to the client or their representative, typically the quantity surveyor or project manager. This claim outlines the work completed to date, quantified and valued according to the contract terms. The process involves several key steps:

- **Measurement and Quantification**: The contractor measures the work completed in the specified period, usually monthly. This includes all activities carried out on-site, such as earthworks, structural work, mechanical and electrical installations, and finishing works. The measurement should be based on the agreed methods, such as the Standard Method of Measurement (SMM) or the New Rules of Measurement (NRM).

- **Valuation**: The measured work is then valued based on the contract rates or agreed unit prices. This valuation should also account for any variations, additional works, or changes in the project scope that have been approved by the client. The valuation should be comprehensive and include all aspects of the work done, ensuring that the contractor is compensated fairly.

- **Submission**: The contractor submits the progress claim to the quantity surveyor or project manager, accompanied by supporting documents such as measurement sheets, work schedules, and any necessary approvals for variations. The submission must adhere to the contract's terms and conditions, including timelines and required documentation.

Payment Applications

Payment applications are the formal documents submitted alongside progress claims, detailing the amount requested for payment. These applications must include a breakdown of the work done, the value of each component, deductions for previous payments, retention amounts, and any other relevant adjustments, such as penalties or incentives. The payment application should be clear and concise, providing all necessary information to facilitate the review and approval process.

Review and Certification

Upon receiving the progress claim and payment application, the quantity surveyor or project manager reviews the documents to verify the accuracy of the measurements, valuations, and compliance with the contract terms. This review includes site inspections, discussions with the contractor, and cross-referencing with the project's progress records. The quantity surveyor ensures that the claimed amounts are justified and align with the work completed.

Once the review is complete, the quantity surveyor or project manager certifies the payment application, approving the amount to be paid. This certification is an official acknowledgment that the claimed work has been completed to the required standard and that the payment is due. The certified payment amount is then submitted to the client for processing, and payment is made to the contractor.

Importance of Accurate and Timely Claims

Accurate and timely progress claims and payment applications are crucial for maintaining the project's financial health. Delays or inaccuracies in these processes can lead to cash flow issues for the contractor, potentially affecting the project's schedule and quality. For the client, prompt and accurate payments help maintain a positive working relationship with the contractor and ensure the project's smooth progression.

Progress claims and payment applications ensure that contractors are compensated for the work completed. Quantity surveyors play a vital role in reviewing and certifying these claims, assuring both the contractor and the client that the payments are fair, accurate, and in accordance with the contract terms. Table 15 outlines common issues encountered in progress claims and payment applications, providing insights into potential challenges and how to address them

Table 15: Issues encountered in progress claims and payment applications

Issue	Description	Mitigation Strategies
Incomplete Documentation	Missing or incomplete supporting documents for claims	Ensure comprehensive documentation
Discrepancies	Differences between claimed amounts and approved budgets	Regular reconciliation and audits
Timing Issues	Delays in submission or processing of claims	Adhere to schedules and deadlines
Calculation Errors	Errors in calculating claimed amounts	Double-check calculations

Accurate Record-Keeping and Documentation

Accurate record-keeping and documentation are pivotal in the effective management of construction projects, particularly from the perspective of quantity surveying. These practices ensure that all financial transactions, progress assessments, and contractual obligations are documented, verified, and available for reference throughout the project lifecycle. Proper documentation supports transparency, accountability, and dispute resolution, thereby contributing to the overall success of the project.

Importance of Accurate Record-Keeping

Accurate record-keeping is crucial for several reasons:

- **Financial Management**: Detailed records of all financial transactions, including progress claims, payment applications, and change orders, are essential for effective financial management. These records help in tracking expenditures, managing budgets, and forecasting future financial needs.

- **Project Progress Monitoring**: Records related to work completed, including measurement and valuation data, enable the tracking of project progress against the schedule. This information is vital for assessing whether the project is on track and for making necessary adjustments to the schedule or resources.

- **Dispute Resolution**: In the event of disputes between parties, comprehensive documentation serves as a critical reference for resolving conflicts. Accurate records of contractual agreements, variations, and communications provide evidence to support claims and counterclaims, facilitating fair and informed resolution.

- **Compliance and Auditing**: Maintaining thorough records ensures compliance with legal, regulatory, and contractual requirements. It also facilitates auditing processes by providing a clear and organized account of all project-related activities.

Types of Documentation

The types of documentation required for accurate record-keeping in construction projects include:

- **Contracts and Agreements**: These documents outline the terms and conditions of the project, including scope, schedule, and payment terms. Contracts should be carefully reviewed and stored for reference throughout the project.

- **Progress Reports**: Regular progress reports document the status of the project, including work completed, issues encountered, and any deviations from the plan. These reports are essential for monitoring progress and communicating with stakeholders.

- **Measurement and Valuation Records**: Detailed records of measurements taken on-site and the corresponding valuations are necessary for progress claims and payment applications. These records should include measurements, unit rates, and any adjustments for variations.

- **Variation Orders**: Documentation of any changes to the original contract, including scope modifications and additional work, is crucial. Variation orders should be detailed, approved by relevant parties, and recorded accurately.

- **Invoices and Receipts**: Invoices and receipts for materials, labor, and other expenses should be carefully documented and matched with progress claims and payment applications to ensure accuracy.

- **Correspondence**: All correspondence related to the project, including emails, letters, and meeting minutes, should be recorded. This documentation provides a record of communications and decisions made throughout the project.

Best Practices for Documentation

To ensure effective record-keeping and documentation, consider the following best practices:

- **Consistency**: Use standardized formats and procedures for documenting all project-related information. This ensures uniformity and ease of retrieval.

- **Accuracy**: Ensure that all records are accurate and up-to-date. Errors or omissions can lead to disputes and financial discrepancies.

- **Organization**: Maintain an organized filing system, both digitally and physically, to facilitate easy access to records. Digital records should be backed up regularly to prevent data loss.

- **Security**: Implement security measures to protect sensitive information from unauthorized access or tampering. This includes both physical and digital security protocols.

- **Timeliness**: Document information as soon as it becomes available. Delayed documentation can lead to inaccuracies and difficulties in managing the project.

Accurate record-keeping and documentation support financial management, project progress monitoring, dispute resolution, and compliance. By adhering to best practices in documentation, quantity surveyors and project managers can ensure that all project-related information is well-organized, accurate, and readily accessible, contributing to the successful execution and completion of the project. Table 16 provides best practices for maintaining accurate records and documentation throughout the construction phase.

Table 16: Maintaining accurate records and documentation

Best Practice	Description	Benefits
Regular Updates	Update records and documentation frequently	Ensures accuracy and completeness
Standardized Formats	Use standardized formats for all documentation	Facilitates consistency and ease of use
Secure Storage	Store records securely, both physically and electronically	Prevents loss or unauthorized access
Periodic Audits	Conduct periodic audits of documentation	Identifies and corrects errors

The Role of Quantity Surveyors in Sustainability and Green

The role of quantity surveyors in sustainability and green building practices is increasingly critical as the construction industry shifts towards more environmentally responsible and resource-efficient practices. Quantity surveyors, traditionally focused on cost management, now play a pivotal role in integrating sustainability principles into construction projects, helping to ensure that environmental considerations are accounted for throughout the project lifecycle.

Understanding Sustainability in Construction

Sustainability in construction refers to practices that minimize the environmental impact of building projects while enhancing social and economic benefits. It encompasses various aspects, including the efficient use of resources, reduction of waste and emissions, and the promotion of environmentally friendly materials and technologies. Green building practices aim to achieve these goals by incorporating sustainable design principles and adhering to recognized standards and certifications.

Integration of Sustainability in Quantity Surveying

Quantity surveyors contribute to sustainability and green building practices in several ways:

- **Cost Planning and Budgeting**: Quantity surveyors help integrate sustainability goals into the project budget by identifying cost-effective green building solutions. They assess the financial implications of sustainable practices and technologies, ensuring that they are economically viable and aligned with the project's budgetary constraints.

- **Material Selection and Specification**: Selecting sustainable materials is a key component of green building. Quantity surveyors guide material choices that have lower environmental impacts, such as recycled, renewable, or low-emission materials. They ensure that material specifications meet sustainability criteria while staying within budget.

- **Life Cycle Cost Analysis**: Life cycle cost analysis evaluates the total cost of a building over its entire lifecycle, including initial construction, operation, maintenance, and disposal. Quantity surveyors use LCCA to assess the long-term economic benefits of sustainable building practices, helping clients make informed decisions about investments in green technologies.

- **Waste Management and Minimization**: Effective waste management is essential for sustainability. Quantity surveyors play a role in developing waste management plans, monitoring construction waste, and implementing strategies to minimize waste generation. They ensure that waste reduction measures are incorporated into the project's cost and scheduling plans.

- **Energy Efficiency**: Enhancing energy efficiency is a major focus of green building. Quantity surveyors assist in evaluating and incorporating energy-efficient systems and technologies, such as high-performance insulation, energy-efficient HVAC systems, and renewable energy sources. They also assess the cost implications and benefits of these systems.

- **Compliance with Green Building Standards**: Quantity surveyors ensure that projects comply with green building standards and certifications, such as LEED (Leadership in Energy and Environmental Design) or BREEAM (Building Research Establishment Environmental Assessment Method). They help prepare documentation required for certification and verify that sustainability criteria are met.

- **Environmental Impact Assessment**: Quantity surveyors participate in environmental impact assessments to evaluate the potential effects of construction projects on the environment. They help identify strategies to mitigate adverse impacts and incorporate sustainable practices into the project's design and execution.

Challenges in Implementing Sustainable Practices

Despite the growing emphasis on sustainability, implementing green building practices can present several challenges:

- **Cost Considerations**: The initial cost of sustainable technologies and materials may be higher than traditional options. Quantity surveyors must balance these costs with the long-term benefits of sustainability and help clients understand the value proposition of green investments.

- **Knowledge and Expertise**: Staying updated with the latest developments in sustainable building practices and technologies requires continuous learning and expertise. Quantity surveyors must invest in acquiring knowledge and skills related to sustainability to effectively contribute to green building projects.

- **Regulatory and Certification Requirements**: Navigating the complexities of green building regulations and certification processes can be challenging. Quantity surveyors must be well-versed in relevant standards and requirements to ensure compliance and successful certification.

- **Stakeholder Engagement**: Engaging stakeholders and ensuring their commitment to sustainability goals can be difficult. Quantity surveyors play a role in communicating the benefits of sustainable practices and aligning stakeholder interests with environmental objectives.

The role of quantity surveyors in sustainability and green building practices is crucial for advancing environmentally responsible construction. By integrating sustainability principles into cost planning, material selection, life cycle cost analysis, and waste management, quantity surveyors help achieve the environmental and economic goals of green building projects. Addressing the challenges associated with sustainable practices requires ongoing expertise and commitment, but the benefits of contributing to a more sustainable built environment are significant. As the construction industry continues to prioritize sustainability, the role of quantity surveyors will remain central to ensuring that green building practices are effectively implemented and managed.

CHAPTER 8:
QUANTITY SURVEYING IN POST-CONSTRUCTION

*Final Account Preparation
and Settlement*

The final account preparation and settlement phase marks the conclusion of a construction project, encapsulating the final financial adjustments between the client and contractor. This phase is crucial as it determines the overall financial outcome of the project, reflecting the total costs and any adjustments to the original contract sum.

The preparation of the final account involves a comprehensive review of all project expenditures, including the initial contract sum, variations, additional works, and any other adjustments that have occurred throughout the construction process. The quantity surveyor plays a pivotal role in this process by ensuring that all financial transactions are accurately recorded and justified. This includes verifying that all change orders, variations, and claims have been appropriately documented and incorporated into the final account.

Key steps in final account preparation include

- **Compilation of Costs**: All incurred costs must be compiled, including labor, materials, and overheads. This involves a detailed reconciliation of the initial contract sum with the actual expenditure incurred.

- **Assessment of Variations**: Any variations or changes to the original scope of work must be assessed and included in the final account. This requires a thorough review of change orders and additional work claims.

- **Resolution of Disputes**: Any disputes regarding costs or changes need to be resolved before the final account can be settled. This may involve negotiations between the client and contractor.

- **Final Account Submission**: Once all adjustments have been made, the final account is prepared and submitted for approval. This document should accurately reflect the total costs and any adjustments agreed upon during the project.

Settlement of the final account typically involves a formal agreement between the client and contractor on the final payment due. Both parties must review and agree on the final account to ensure that all financial matters are concluded satisfactorily. Any outstanding issues should be addressed before final payment is made, as this can help prevent future disputes.

The final account not only concludes the financial aspects of the project but also serves as a crucial document for financial reporting and auditing purposes. It provides a clear record of the project's financial performance and serves as a reference for any future projects.

Cost Control and Project Performance Analysis

Cost control and project performance analysis are critical components in the post-construction phase of a project. This stage involves evaluating the financial management of the project and analyzing the effectiveness of the cost control measures implemented throughout its lifecycle.

Cost Control

Effective cost control throughout the project lifecycle ensures that the project remains within its budget and that any deviations are managed promptly. In the post-construction phase, cost control involves:

- **Reviewing Final Costs**: Analyzing the final costs against the original budget helps identify any discrepancies. This review includes assessing direct costs, indirect costs, and any adjustments due to changes or unforeseen circumstances.

- **Assessing Variations and Claims**: Evaluating the impact of variations and claims on the final budget is essential. It involves reviewing the reasons for changes and determining whether they were justified and managed effectively.

- **Identifying Cost Savings and Efficiencies**: Post-project analysis provides an opportunity to identify areas where cost savings were achieved or where efficiencies could have been improved. This information is valuable for future projects and for refining cost control strategies.

Project Performance Analysis

Project performance analysis involves evaluating how well the project met its financial and performance objectives. Key aspects include:

- **Budget Adherence**: Assessing whether the project adhered to its budget helps evaluate the effectiveness of the financial planning and management processes. This includes reviewing any budget overruns and the reasons behind them.

- **Schedule Performance**: Analyzing whether the project was completed on time and the impact of any delays on the overall cost. Delays can lead to increased costs due to extended labor and overheads.

- **Quality of Work**: Evaluating the quality of the completed work against the project's specifications and standards. This analysis helps assess whether any additional costs were incurred due to rework or quality issues.

- **Stakeholder Satisfaction**: Gathering feedback from stakeholders, including clients, contractors, and project managers, to assess their satisfaction with the project outcomes. This feedback can provide insights into areas for improvement and lessons learned.

This analysis helps in understanding the effectiveness of the cost management strategies employed and provides a basis for improving practices in future projects.

Post-Project Reviews and Lessons Learned

Post-project reviews and the extraction of lessons learned are vital components of the project lifecycle, focusing on reflection and improvement. This phase is crucial for identifying what went well, what could be improved, and how future projects can benefit from these insights.

Post-Project Reviews

Post-project reviews, also known as project close-out meetings or post-mortem analyses, involve a comprehensive evaluation of the project once it is completed. The primary objectives of these reviews are to:

- **Assess Project Outcomes**: Review whether the project met its objectives, including scope, budget, and schedule. This assessment helps determine the overall success of the project and provides a basis for evaluating performance.

- **Evaluate Financial Performance**: Analyze the final financial performance against the initial budget and financial forecasts. This includes reviewing any cost overruns, savings, and financial management practices.

- **Review Project Management Processes**: Evaluate the effectiveness of the project management processes and methodologies used. This review includes assessing planning, execution, monitoring, and control processes.

- **Gather Stakeholder Feedback**: Collect feedback from

all stakeholders involved in the project, including clients, contractors, team members, and suppliers. This feedback provides insights into the project's impact and identifies areas for improvement.

Lessons Learned

The lessons learned process involves documenting and analyzing the knowledge gained from the project. This process includes:

- **Identifying Key Lessons**: Determine the critical lessons learned from various aspects of the project, such as planning, execution, cost management, and stakeholder engagement. This identification involves reviewing both successes and challenges encountered during the project.

- **Documenting Best Practices**: Record best practices and effective strategies that contributed to the project's success. This documentation serves as a reference for future projects and helps in replicating successful approaches.

- **Recognizing Areas for Improvement**: Highlight areas where improvements are needed, such as inefficiencies, process shortcomings, or communication gaps. Understanding these areas helps in addressing similar issues in future projects.

- **Updating Organizational Knowledge**: Integrate the lessons learned into the organization's knowledge base and project management practices. This integration ensures that the insights gained are accessible and utilized for future projects.

The Importance of Post-Project Reviews and Lessons Learned

Post-project reviews and lessons learned are essential for continuous improvement and organizational learning. They provide valuable feedback that enhances project management practices, improves decision-making, and contributes to the overall success of future projects. By systematically analyzing project outcomes and documenting lessons learned, organizations can build a foundation for better project execution and management.

Tips for Effective Communication During Final Account Settlement

Effective communication during the final account settlement phase is crucial for ensuring clarity, avoiding disputes, and achieving a smooth conclusion to the project. This stage involves negotiating and finalizing the final account, which includes the settlement of all financial matters related to the project. To facilitate a successful final account settlement, consider the following tips:

1. Establish Clear Communication Channels

At the outset of the final account settlement process, establish and agree upon communication channels with all parties involved. This includes defining the preferred methods for communication (e.g., email, meetings, written reports) and identifying key contacts for each party. Clear communication channels help ensure that information is accurately conveyed and that responses are timely.

2. Provide Comprehensive Documentation

Ensure that all relevant documentation is complete, accurate, and readily accessible. This includes contracts, change orders, progress reports, and any other documents related to financial transactions. Comprehensive documentation supports transparency and facilitates the resolution of any discrepancies that may arise during the settlement process.

3. Maintain Detailed Records

Keep detailed records of all communications related to the final account settlement. This includes written correspondence, meeting minutes, and any agreements or amendments. Detailed records serve as a reference and can help resolve any misunderstandings or disputes that may occur.

4. Schedule Regular Meetings

Arrange regular meetings with all relevant stakeholders to review progress and address any issues related to the final account settlement. These meetings should be structured to cover key topics, such as outstanding payments, adjustments, and any potential disputes. Regular meetings help maintain momentum and ensure that all parties are aligned on the status of the settlement.

5. Clearly Articulate Financial Adjustments

When discussing financial adjustments or changes to the final account, clearly articulate the reasons for these adjustments and provide supporting evidence. This includes detailing any additional costs, deductions, or changes to the scope of work. Clear articulation helps in minimizing misunderstandings and facilitates a smoother negotiation process.

6. Address Disputes Promptly

In the event of any disputes or disagreements, address them promptly and professionally. Engage in constructive dialogue to understand the concerns of all parties and seek to reach a fair resolution. Timely resolution of disputes helps in avoiding delays and ensuring that the final account settlement is completed efficiently.

7. Ensure Transparency

Foster transparency throughout the final account settlement process by openly sharing information and updates with all parties. Transparency builds trust and helps in maintaining positive working relationships. It also reduces the likelihood of disputes arising from misunderstandings or lack of information.

8. Document Agreements Thoroughly

Once an agreement is reached on the final account, document it thoroughly and ensure that all parties sign off on the final settlement. This documentation serves as a formal record of the agreed terms and provides a basis for resolving any future issues that may arise.

9. Review and Confirm

Before finalizing the account, review all calculations, adjustments, and agreements to ensure accuracy. Confirm that all parties are in agreement with the final figures and terms. A thorough review helps prevent errors and ensures that the final account settlement is accurate and complete.

10. Provide a Final Summary Report

After the settlement process, provide a final summary report that outlines the key outcomes and agreements. This report should be clear and concise, summarizing the final account details and any relevant information. A final summary report serves as a useful reference and formalizes the conclusion of the settlement process.

Effective communication during the final account settlement phase is essential for achieving a successful and amicable resolution. By following these tips, quantity surveyors can facilitate a smooth settlement process, minimize disputes, and ensure that all financial matters are resolved satisfactorily.

PART 4: ESSENTIAL SKILLS FOR QUANTITY SURVEYORS

CHAPTER 9: QUANTITY SURVEYING AND SUSTAINABILITY

Introduction to Sustainable Construction Practices

Sustainable construction practices have become a cornerstone in the modern construction industry, driven by the increasing need to minimize environmental impacts and promote long-term economic and social benefits. The concept of sustainability in construction encompasses a broad range of practices that aim to reduce the carbon footprint of buildings, conserve natural resources, and enhance the quality of life for present and future generations.

At the core of sustainable construction is the principle of environmental stewardship. This involves the careful management of natural resources, such as water, energy, and raw materials, to ensure their availability for future use. Sustainable construction practices emphasize the use of renewable resources and the minimization of waste and pollution. For example, the selection of sustainable building materials, such as recycled steel or reclaimed wood, can significantly reduce the environmental impact of construction projects.

Another critical aspect of sustainable construction is energy efficiency. Buildings are major consumers of energy, and improving their energy performance is a key focus of sustainability efforts. This can be achieved through various strategies, including the design of energy-efficient building envelopes, the use of energy-efficient systems and appliances, and the incorporation of renewable energy sources, such as solar panels and wind turbines. Energy-efficient buildings not only reduce greenhouse gas emissions but also offer significant cost savings over their operational life.

Water conservation is another vital component of sustainable construction. This includes the efficient use of water resources during construction and throughout the building's lifecycle. Techniques such as rainwater harvesting, greywater recycling, and the use of low-flow fixtures can help reduce water consumption. Additionally, landscaping practices that utilize native and drought-resistant plants can minimize the need for irrigation.

Indoor environmental quality is also a key consideration in sustainable construction. A healthy indoor environment is essential for the well-being of occupants and can be achieved through the use of non-toxic building materials, adequate ventilation, and natural lighting. Sustainable buildings prioritize the use of materials that emit low levels of volatile organic compounds and other pollutants. Moreover, incorporating features that maximize natural light and fresh air contributes to a healthier and more comfortable indoor environment.

In addition to environmental considerations, sustainable construction practices also address economic and social dimensions. Economically, sustainable buildings often have lower operating costs due to reduced energy and water consumption, and they can command higher market values and rental rates. Socially, sustainable construction contributes to the well-being of communities by providing healthier living and working environments, promoting social equity, and supporting local economies through the use of local materials and labor.

One of the frameworks widely adopted in promoting sustainable construction is the use of green building rating systems. These systems, such as Leadership in Energy and Environmental Design, Building Research Establishment Environmental Assessment Method, and Green Star, provide guidelines and standards for designing, constructing, and operating buildings in a sustainable manner. They assess various aspects of a building's performance, including energy efficiency, water use, materials selection, and indoor environmental quality, awarding certifications based on the building's overall sustainability performance.

The transition to sustainable construction practices requires a collaborative effort from all stakeholders, including architects, engineers, contractors, developers, and policymakers. It involves rethinking traditional construction methods and embracing innovative approaches and technologies. For example, the use of Building Information Modeling (BIM) can enhance the planning and design process, enabling more accurate assessments of a building's environmental impact and performance. Additionally, advancements in materials science, such as the development of high-performance insulation and energy-efficient glazing, play a crucial role in achieving sustainability goals.

Sustainable construction practices are essential for addressing the environmental, economic, and social challenges facing the construction industry today. By prioritizing resource efficiency, reducing emissions, and enhancing the quality of life, these practices pave the way for a more sustainable and resilient built environment. As the demand for sustainable buildings continues to grow, quantity surveyors play a vital role in integrating sustainability into construction projects, from cost estimation and budgeting to procurement and project management. Their expertise is crucial in ensuring that sustainability goals are met within the constraints of time, budget, and quality.

The Role of Quantity Surveyors in Promoting Sustainability

Quantity surveyors play a pivotal role in promoting sustainability within the construction industry. Their expertise in cost management, procurement, and project management uniquely positions them to influence sustainable practices throughout the lifecycle of a construction project. By integrating sustainability into their core responsibilities, quantity surveyors contribute to the development of environmentally responsible, economically viable, and socially equitable buildings.

One of the primary roles of quantity surveyors in promoting sustainability is through cost planning and feasibility studies. At the early stages of a project, quantity surveyors can assess the financial implications of incorporating sustainable practices and materials. They provide clients with detailed cost estimates that highlight the potential long-term savings and benefits of investing in energy-efficient systems, renewable energy sources, and sustainable materials. By presenting a comprehensive cost-benefit analysis, quantity surveyors help clients make informed decisions that balance initial capital expenditure with future operational savings and environmental impact.

In the procurement process, quantity surveyors ensure that sustainability criteria are included in the selection of contractors, suppliers, and materials. They can specify the use of environmentally friendly materials, such as low-VOC paints, recycled steel, and sustainably sourced timber, in the tender documents. Furthermore, quantity surveyors can advocate for the inclusion of sustainability clauses in contracts, which mandate the use of green building practices and technologies. By setting these requirements, quantity surveyors drive the adoption of sustainable construction methods and materials throughout the supply chain.

During the construction phase, quantity surveyors monitor and manage the implementation of sustainable practices. They ensure that contractors adhere to specified sustainability standards and that any deviations are promptly addressed. Quantity surveyors also track the performance of sustainable systems and materials, ensuring they meet the expected energy efficiency, water conservation, and waste reduction targets. Their role in project management allows them to coordinate with other stakeholders, such as architects, engineers, and environmental consultants, to ensure a cohesive approach to sustainability.

Another critical aspect of promoting sustainability is lifecycle costing, a process in which quantity surveyors evaluate the total cost of ownership of a building, including construction, operation, maintenance, and eventual disposal. Lifecycle costing helps identify opportunities for cost savings through the use of durable, low-maintenance materials and energy-efficient systems. By considering the long-term financial and environmental impacts, quantity surveyors can advocate for sustainable solutions that may have higher upfront costs but result in significant savings and reduced environmental impact over the building's lifespan.

Quantity surveyors also play a role in promoting sustainability through their involvement in certification processes for green building rating systems, such as LEED, BREEAM, and Green Star. They gather and compile the necessary documentation to demonstrate compliance with the sustainability criteria set by these rating systems. This includes tracking and reporting on the use of sustainable materials, energy and water efficiency measures, and indoor environmental quality improvements. By facilitating the certification process, quantity surveyors help projects achieve recognition for their sustainable efforts, which can enhance the building's marketability and value.

Education and advocacy are additional ways in which quantity surveyors promote sustainability. They stay informed about the latest advancements in sustainable construction practices, materials, and technologies, and they share this knowledge with clients and colleagues. By promoting a culture of sustainability within their organizations and the broader construction industry, quantity surveyors help raise awareness and drive the adoption of best practices. They can also participate in industry forums, conferences, and working groups focused on sustainability, contributing to the development of standards and guidelines that support sustainable construction.

Cost-Benefit Analysis of Green Building Practices

The cost-benefit analysis of green building practices is a critical aspect of decision-making in sustainable construction. This analysis involves evaluating the economic, environmental, and social benefits of implementing sustainable practices against the associated costs. By conducting a thorough cost-benefit analysis, quantity surveyors and other stakeholders can make informed decisions that balance the financial investment with long-term gains, ensuring the feasibility and desirability of green building initiatives.

At the core of the cost-benefit analysis is the initial cost of implementing green building practices. These costs can include higher expenditures on sustainable materials, energy-efficient systems, and technologies such as solar panels, rainwater harvesting systems, and advanced HVAC systems. While these upfront costs may be higher compared to conventional building practices, the analysis must consider the potential savings over the building's lifecycle. For instance, energy-efficient systems can significantly reduce operational costs by lowering energy consumption, resulting in substantial savings on utility bills over time.

One of the primary benefits of green building practices is the reduction in operating costs. Energy-efficient buildings consume less electricity and water, leading to lower utility expenses. For example, the use of energy-efficient lighting, heating, and cooling systems can reduce energy consumption by a significant percentage. Additionally, the incorporation of water-saving fixtures and systems can decrease water usage, which is particularly valuable in regions with water scarcity. These operational savings often offset the initial investment in green technologies, making the building more cost-effective over its lifespan.

Another key benefit of green building practices is the enhancement of indoor environmental quality (IEQ). Green buildings prioritize the use of non-toxic, low-emission materials, proper ventilation, and natural lighting, which contribute to a healthier indoor environment. Improved IEQ can lead to better health and productivity outcomes for occupants, which is especially important in commercial and institutional buildings. For employers, this can translate into reduced absenteeism, increased employee satisfaction, and higher productivity levels. Although these benefits may be challenging to quantify directly, they add significant value to the overall cost-benefit equation.

Green building practices also contribute to environmental sustainability by reducing the carbon footprint and conserving natural resources. The use of renewable energy sources, such as solar and wind, decreases reliance on fossil fuels and reduces greenhouse gas emissions. Additionally, sustainable materials often have lower embodied energy, meaning less energy is consumed during their production, transportation, and installation. This reduction in carbon emissions is a critical factor in addressing climate change and meeting global sustainability goals. The environmental benefits, while not always immediately reflected in financial terms, contribute to the long-term sustainability and resilience of the built environment.

The marketability and value of green buildings are other important considerations in the cost-benefit analysis. Buildings certified under green building rating systems, such as LEED, BREEAM, or Green Star, often command higher rental and sale prices compared to conventional buildings. This premium is attributed to the growing demand for sustainable properties, as tenants and buyers increasingly prioritize environmentally responsible buildings. Moreover, green buildings tend to have lower vacancy rates and higher tenant retention, further enhancing their financial performance. The certification process itself, while incurring costs, can be seen as an investment that yields higher returns through increased property value and market appeal.

Furthermore, green building practices can reduce regulatory and compliance risks. Many governments and local authorities are introducing stricter building codes and regulations related to energy efficiency, water conservation, and waste management. By adopting green practices, developers and building owners can ensure compliance with these evolving standards, avoiding potential fines, penalties, and costly retrofits. Proactively implementing sustainable practices can also position the building as a leader in environmental stewardship, enhancing its reputation and brand image.

In the context of social benefits, green building practices contribute to the well-being of communities and the broader society. Sustainable buildings often include features such as green roofs, public spaces, and community gardens, which enhance the quality of life for residents and visitors. These features can promote social interaction, physical activity, and mental well-being. Additionally, green buildings that incorporate accessibility features and are designed to be inclusive contribute to social equity, ensuring that all individuals, regardless of their abilities, can benefit from the built environment.

Case Studies on Sustainable Projects

The examination of case studies on sustainable projects provides invaluable insights into the practical application of green building principles and the role of quantity surveyors in promoting sustainability. These case studies highlight the challenges, solutions, and outcomes of implementing sustainable practices, offering lessons that can guide future projects. They demonstrate how innovative design, material selection, and construction methods contribute to sustainability goals while maintaining economic viability and enhancing occupant well-being.

Case Study 1: The Edge, Amsterdam

The Edge, located in Amsterdam, is often cited as one of the most sustainable office buildings in the world. Designed by PLP Architecture and developed by OVG Real Estate, The Edge showcases how advanced technology and sustainable design can be integrated to create a highly efficient and comfortable workspace. The building achieved a BREEAM Outstanding rating, the highest possible score, due to its innovative features and sustainable strategies.

Key features of The Edge include an extensive use of solar panels, a highly efficient climate control system, and smart building technology that optimizes energy use. The building's design maximizes natural light, reducing the need for artificial lighting and thereby lowering energy consumption. Additionally, rainwater is harvested and used for flushing toilets and irrigating green areas, further reducing the building's environmental footprint.

The role of quantity surveyors in this project involved the careful evaluation of the cost implications of sustainable features and ensuring that the project remained within budget while achieving its sustainability targets. This case study demonstrates the importance of integrating sustainability considerations early in the project planning phase and highlights the economic and environmental benefits of investing in sustainable technologies.

Case Study 2: One Central Park, Sydney
One Central Park in Sydney, Australia, is another exemplary sustainable project that combines residential, commercial, and retail spaces. Designed by Ateliers Jean Nouvel and PTW Architects, this development is renowned for its vertical gardens and integrated green technologies. The building has received numerous awards, including the Best Tall Building Worldwide from the Council on Tall Buildings and Urban Habitat (CTBUH).

The vertical gardens, designed by botanist Patrick Blanc, cover a significant portion of the building's façade, enhancing its aesthetic appeal and providing natural insulation. The development also includes a central thermal plant, which provides heating and cooling through a highly efficient tri-generation system. This system uses natural gas to generate electricity, with the byproduct heat used for hot water and space heating, thus maximizing energy efficiency.

Quantity surveyors played a crucial role in the One Central Park project by conducting life cycle cost analyses to evaluate the long-term savings from reduced energy and maintenance costs. They also facilitated value engineering processes to optimize the design and ensure that sustainability measures were cost-effective. This case study underscores the importance of innovative design in achieving sustainability and the value of thorough cost assessments in decision-making.

Case Study 3: The Crystal, London
The Crystal in London, a project developed by Siemens, is a sustainable building dedicated to showcasing the latest in sustainable urban technologies. The building achieved both BREEAM Outstanding and LEED Platinum certifications, reflecting its exceptional sustainability performance. The Crystal serves as a hub for dialogue and research on sustainable urban development, housing exhibition spaces and conference facilities.

Key sustainable features of The Crystal include the use of ground source heat pumps, solar panels, and rainwater harvesting systems. The building's design emphasizes energy efficiency, with triple-glazed windows and advanced building management systems that monitor and optimize energy use. The Crystal also incorporates sustainable transportation solutions, such as electric vehicle charging stations and bicycle facilities, promoting low-carbon mobility.

In the context of this project, quantity surveyors were involved in assessing the financial feasibility of incorporating various sustainable technologies. They provided detailed cost estimates and conducted financial analyses to demonstrate the long-term benefits of investing in energy-efficient systems. The Crystal exemplifies how sustainable buildings can serve as educational and inspirational tools, promoting awareness and innovation in sustainable practices.

Case Study 4: Bosco Verticale, Milan

Bosco Verticale, or "Vertical Forest," in Milan, Italy, is a residential tower known for its extensive use of greenery integrated into the building structure. Designed by Stefano Boeri Architetti, this project consists of two towers covered with approximately 20,000 trees and plants. The vegetation is carefully selected to provide natural shading, improve air quality, and enhance biodiversity within the urban environment.

The incorporation of greenery in Bosco Verticale serves multiple purposes, including reducing the urban heat island effect, providing insulation, and creating a natural habitat for birds and insects. The project's design also includes renewable energy systems, such as solar panels, to power common areas and reduce overall energy consumption.

Quantity surveyors involved in Bosco Verticale focused on managing the costs associated with the innovative use of vegetation and renewable energy systems. They conducted thorough market analyses to source cost-effective materials and technologies, ensuring that the project remained within budget while achieving its environmental goals. This case study highlights the role of creative design in urban sustainability and the importance of detailed cost planning in managing innovative projects.

These case studies illustrate the diverse approaches and strategies employed in sustainable construction projects around the world. They highlight the critical role of quantity surveyors in not only managing costs but also ensuring that sustainability objectives are met. Through careful planning, cost analysis, and innovative design, these projects demonstrate that sustainable building practices can be both environmentally responsible and economically viable. They serve as valuable examples for future projects, offering insights into best practices and the potential benefits of integrating sustainability into the construction industry. Table 17 presents a summary of the sustainable features and the roles of quantity surveyors in each case study.

Table 17: Case study

Case Study	Sustainable Features	Quantity Surveyor's Role
The Edge, Amsterdam	Solar panels, efficient climate control, smart building technology, natural light maximization, rainwater harvesting	Evaluated cost implications of sustainable features, ensured budget adherence
One Central Park, Sydney	Vertical gardens, central thermal plant with tri-generation system	Conducted life cycle cost analyses, facilitated value engineering
The Crystal, London	Ground source heat pumps, solar panels, rainwater harvesting, sustainable transportation solutions	Assessed financial feasibility of sustainable technologies, provided detailed cost estimates, conducted financial analyses
Bosco Verticale, Milan	Extensive use of greenery, improved air quality, enhanced biodiversity, renewable energy systems	Managed costs associated with vegetation and renewable systems, conducted market analyses for cost-effective materials

CHAPTER 10: DIGITAL TECHNOLOGIES IN QUANTITY SURVEYING

Overview of Digital Tools and
Software in Quantity Surveying

The integration of digital tools and software in quantity surveying has revolutionized the profession, enhancing accuracy, efficiency, and collaboration. Traditionally, quantity surveyors relied heavily on manual methods and paper-based documentation, which were time-consuming and prone to errors. The advent of digital technologies has introduced a range of tools that streamline processes, improve data management, and facilitate more precise measurement and estimation.

One of the most significant advancements in this field is the use of specialized software for cost estimation, measurement, and project management. Software like CostX, Bluebeam, and PlanSwift offer functionalities that allow quantity surveyors to take off quantities directly from digital drawings, generate detailed cost estimates, and produce accurate bills of quantities (BOQs). These tools are equipped with features such as digital plan takeoff, which enables the extraction of quantities from digital drawings with high precision, thus minimizing the risk of measurement errors.

Furthermore, digital tools facilitate the integration of data from various sources, enabling a holistic approach to project cost management. For example, software solutions often include modules for project scheduling, risk analysis, and financial reporting, which are critical for comprehensive project planning and execution. The ability to integrate these aspects into a unified platform allows quantity surveyors to manage project data more effectively, ensuring that all stakeholders have access to consistent and up-to-date information.

Another critical area where digital tools have made a significant impact is in documentation and record-keeping. Electronic document management systems have replaced traditional filing systems, allowing for better organization and retrieval of documents. These systems offer features like version control, audit trails, and secure access, which are essential for maintaining the integrity and confidentiality of project data. This digital shift not only enhances productivity but also supports compliance with industry regulations and standards.

The adoption of digital tools in quantity surveying is not without challenges. The initial cost of software acquisition and training can be high, and there may be resistance to change among professionals accustomed to traditional methods. However, the long-term benefits, including improved accuracy, reduced project timelines, and enhanced collaboration, far outweigh the initial investment. The increasing complexity of construction projects and the demand for more sustainable and efficient building practices make the adoption of digital technologies not just advantageous but necessary. Table 18 presents an overview of key digital tools and software commonly used in quantity surveying.

Table 18: Overview of Key Digital Tools and Software in Quantity Surveying

Tool/Software	Key Features	Applications in Quantity Surveying	Examples
Building Information Modeling	3D modeling, data integration, collaboration	Design visualization, cost estimation, project management	Revit, ArchiCAD, Bentley Systems
Cost Estimating Software	Automated takeoff, cost database, reporting	Preparing cost estimates, budgeting, bid preparation	CostX, Sage Estimating, PlanSwift
Project Management Software	Scheduling, resource management, progress tracking	Project scheduling, risk management, progress monitoring	Primavera, MS Project, Asta Powerproject
Measurement and Takeoff Tools	Digital measurements, plan analysis, quantity takeoff	Accurate quantity measurements, efficient takeoff process	Bluebeam Revu, On-Screen Takeoff
Document Management Systems	Document storage, version control, secure access	Managing project documents, collaboration, compliance	Procore, Viewpoint, Aconex
Blockchain Technology	Decentralized ledger, smart contracts, transparency	Secure transactions, contract management, payment automation	Ethereum, Hyperledger Fabric
Augmented Reality and Virtual Reality	Interactive visualization, immersive experience	Design review, client presentations, site walkthroughs	HoloLens, Oculus Rift, HTC Vive
Internet of Things	Real-time data collection, sensor integration	Building performance monitoring, predictive maintenance	Smart thermostats, energy meters
Big Data Analytics	Data analysis, pattern recognition, predictive modeling	Cost analysis, project benchmarking, risk assessment	

Building Information Modeling and Its Applications

Building Information Modeling is a transformative tool in the construction industry, fundamentally altering how projects are designed, constructed, and managed. For quantity surveyors, BIM offers significant advantages, providing a comprehensive digital representation of a building's physical and functional characteristics. This technology facilitates greater collaboration among stakeholders and enhances the accuracy and efficiency of project management tasks.

At its core, BIM is a process that involves creating and managing digital models of a building. These models are rich with data and can encompass everything from architectural and structural components to mechanical, electrical, and plumbing systems. The integration of these various elements into a single cohesive model allows for a more holistic view of the project, improving coordination and reducing the likelihood of errors or omissions.

For quantity surveyors, one of the most significant applications of BIM is in the area of quantity takeoff and cost estimation. Traditional methods often require manual measurement from 2D drawings, which can be time-consuming and prone to errors. BIM, on the other hand, allows for automatic extraction of quantities from the 3D model. This not only increases accuracy but also saves time, enabling quantity surveyors to focus on more complex aspects of cost management and analysis.

Furthermore, BIM facilitates better visualization of the project, making it easier to identify potential issues and discrepancies at an early stage. This is particularly valuable in clash detection, where different building systems are checked for potential conflicts. For example, BIM can highlight instances where ductwork might interfere with structural elements, allowing for adjustments to be made before construction begins. This proactive approach can significantly reduce the need for costly changes and rework during the construction phase.

BIM also enhances collaboration among project stakeholders, including architects, engineers, contractors, and quantity surveyors. The shared digital model serves as a single source of truth, ensuring that all parties are working from the same information. This integrated approach promotes better communication and decision-making, reducing the likelihood of misunderstandings or misinterpretations. For quantity surveyors, this means more reliable cost estimates and a clearer understanding of project requirements.

In addition to its use in design and construction, BIM has

applications in facility management and lifecycle costing. The data-rich nature of BIM models allows for detailed tracking of building components and systems over the building's lifespan. This information is invaluable for maintenance planning, asset management, and long-term cost forecasting. For quantity surveyors, this means a more comprehensive approach to cost management, encompassing not just initial construction costs but also ongoing operational expenses.

Despite its many advantages, the adoption of BIM is not without challenges. It requires a significant investment in software, training, and process changes. Moreover, the successful implementation of BIM relies on a collaborative culture and a willingness to share data among all project stakeholders. There may also be issues related to data security and intellectual property, particularly concerning the sharing of sensitive information.

The Impact of Digital Technologies on Measurement and Estimating

The advent of digital technologies has revolutionized measurement and estimating processes in the construction industry, bringing unprecedented accuracy, efficiency, and consistency. For quantity surveyors, these technologies offer transformative tools that streamline tasks, reduce errors, and enhance the overall quality of their work. This section explores the significant impact of digital technologies on measurement and estimating practices.

One of the most notable advancements is the development of digital measurement software. These tools allow for precise and automated measurement of quantities from digital drawings and models. Unlike traditional manual methods, which can be labor-intensive and prone to human error, digital tools provide a more accurate and efficient means of capturing data. They enable quantity surveyors to quickly generate accurate measurements for materials, labor, and equipment, directly from digital files. This automation not only saves time but also reduces the potential for mistakes that can arise from manual calculations.

Building Information Modeling stands at the forefront of digital transformation in measurement and estimating. BIM integrates various aspects of construction projects into a unified digital model, providing a comprehensive view of the project's scope and details. This integration allows for more accurate and dynamic quantity takeoffs. BIM models contain detailed information about every component of the building, from structural elements to finishes, enabling precise extraction of quantities. This level of detail facilitates better cost estimation and more informed decision-making throughout the project lifecycle.

Furthermore, digital technologies have introduced advancements such as laser scanning and photogrammetry, which have significantly improved site measurement accuracy. Laser scanning uses laser beams to capture the physical dimensions of a space, creating a highly accurate 3D model. This technology is particularly useful for as-built surveys, where precise measurements are crucial for renovation or retrofitting projects. Similarly, photogrammetry uses photographic images to reconstruct three-dimensional models, providing another layer of accuracy and detail. These technologies enhance the ability of quantity surveyors to measure existing conditions accurately and integrate this data into the estimating process.

The use of digital tools also facilitates better collaboration and communication among project stakeholders. Digital platforms allow for real-time sharing and updating of project data, ensuring that all parties have access to the latest information. This transparency reduces the likelihood of discrepancies and misunderstandings, leading to more accurate and consistent estimates. Moreover, digital platforms often include version control features, which help track changes and revisions, providing a clear audit trail that is essential for accountability and quality control.

Another significant impact of digital technologies is the improvement in data analysis capabilities. Advanced software tools can analyze large datasets to identify trends, patterns, and anomalies, providing valuable insights for cost estimation and risk management. For example, predictive analytics can forecast potential cost overruns based on historical data and current project conditions, allowing for proactive adjustments to estimates and budgets. This data-driven approach enhances the accuracy and reliability of cost estimates, contributing to better project outcomes.

Despite these advancements, the adoption of digital technologies in measurement and estimating is not without challenges. The initial investment in software and training can be substantial, and there may be a learning curve associated with new tools. Additionally, the integration of digital technologies requires a cultural shift within organizations, as traditional methods and mindsets must adapt to new workflows and processes. Data security and intellectual property concerns also arise, particularly when sharing sensitive project information across digital platforms.

Digital technologies have profoundly impacted measurement and estimating in quantity surveying, offering numerous benefits in terms of accuracy, efficiency, and collaboration. Tools such as digital measurement software, BIM, laser scanning, and photogrammetry have transformed traditional practices, enabling more precise and dynamic quantity takeoffs and cost estimates. As these technologies continue to evolve, they will further enhance the capabilities of quantity surveyors, enabling them to deliver more accurate and reliable estimates, manage risks more effectively, and contribute to the overall success of construction projects.

Future Trends in Digital Quantity Surveying

The field of quantity surveying is undergoing a transformative shift, driven by the rapid advancement of digital technologies. These innovations are not only enhancing the efficiency and accuracy of traditional practices but are also paving the way for new methodologies and approaches. This section explores the future trends in digital quantity surveying, focusing on emerging technologies and their potential to revolutionize the profession.

One of the most significant future trends is the continued development and integration of Building Information Modeling. As BIM technology evolves, it is expected to become even more comprehensive and accessible. The future will likely see BIM being used not just for design and construction phases but also for facilities management and operations. This expansion will enable quantity surveyors to provide more holistic services, including life cycle costing and maintenance planning. Advanced BIM models will include detailed data on building components, performance metrics, and maintenance schedules, facilitating better decision-making throughout the building's life span.

Another promising trend is the rise of artificial intelligence and machine learning in quantity surveying. These technologies have the potential to revolutionize data analysis and predictive modeling. AI algorithms can analyze vast amounts of data from past projects to identify patterns and predict future outcomes, such as cost overruns or delays. Machine learning can continuously improve these predictions by learning from new data, making the estimating process more accurate over time. Additionally, AI can automate repetitive tasks, such as data entry and report generation, freeing up quantity surveyors to focus on more strategic and analytical aspects of their work.

The adoption of blockchain technology is another trend poised to impact the construction industry, including quantity surveying. Blockchain provides a decentralized and secure way of recording transactions, which can enhance transparency and trust in construction projects. For quantity surveyors, blockchain can streamline the contract administration process by automating payments and ensuring that all parties have access to the same, immutable data. This technology can also reduce disputes over payment and work done, as the blockchain ledger provides a clear and indisputable record of transactions and agreements.

Augmented reality and virtual reality are also expected to play a more significant role in quantity surveying. These immersive technologies can provide a more intuitive and interactive way of visualizing construction projects. AR can overlay digital information onto the physical world, allowing quantity surveyors to see how changes or additions will look in real life before they are made. This capability can be particularly useful in client presentations and stakeholder meetings, providing a clear and tangible understanding of proposed designs and changes. VR, on the other hand, can create a fully immersive experience, enabling quantity surveyors to explore virtual models of buildings and assess various design options in a simulated environment.

The Internet of Things is another emerging technology that holds promise for quantity surveying. IoT involves the use of sensors and connected devices to collect real-time data on various aspects of a building, such as energy usage, temperature, and occupancy levels. This data can be invaluable for quantity surveyors in conducting post-occupancy evaluations and cost analysis. For instance, real-time data on energy consumption can help in assessing the effectiveness of energy-saving measures and identifying areas for improvement. IoT can also facilitate predictive maintenance by providing early warnings of potential equipment failures, allowing for timely interventions that can save costs.

Finally, the future of digital quantity surveying will likely involve a greater emphasis on data analytics and big data. As the construction industry becomes more data-driven, quantity surveyors will need to develop skills in data analysis and interpretation. The ability to analyze large datasets and extract meaningful insights will be crucial for making informed decisions and providing accurate estimates. Big data analytics can also help in benchmarking projects, identifying trends, and optimizing resource allocation.

CHAPTER 11:
COMMUNICATION
AND
COLLABORATION
SKILLS

Effective Communication with Project Stakeholders

E ffective communication with project stakeholders is a cornerstone of successful project management and quantity surveying. It encompasses the exchange of information among all parties involved in a construction project, including clients, contractors, consultants, and regulatory bodies. The primary objective is to ensure that all stakeholders have a clear understanding of project requirements, progress, and any issues that may arise. This level of communication helps in managing expectations, mitigating conflicts, and fostering a collaborative environment.

The communication process should be structured and systematic. Initially, establishing a communication plan is essential. This plan outlines the communication objectives, identifies key stakeholders, and determines the frequency and methods of communication. It is crucial to use various communication channels effectively, including formal reports, meetings, emails, and informal discussions. Each method serves a different purpose and audience. For example, formal reports and meetings are typically used for major updates and strategic discussions, whereas emails and informal conversations may address day-to-day operational issues.

Active listening plays a significant role in effective communication. It involves not only hearing the words spoken by stakeholders but also understanding their underlying concerns and motivations. This approach helps in accurately interpreting their needs and expectations, leading to better decision-making and problem-solving. Furthermore, feedback mechanisms should be integrated into the communication process. Regularly seeking and providing feedback ensures that communication remains a two-way street, allowing for continuous improvement and alignment with stakeholder expectations.

Transparency is another critical aspect of effective communication. Providing clear and honest information about project status, challenges, and changes builds trust among stakeholders. It is important to present data and updates in a way that is accessible and understandable to all parties involved. This can be achieved by using visual aids such as charts and graphs, which help convey complex information more comprehensively. Table 19 summarizes various communication techniques, their characteristics, and their impact on stakeholder relationships and project success.

Table 19: Communication techniques

Communication Technique	Description	Characteristics	Impact on Stakeholder Relationships	Impact on Project Success
Active Listening	Fully concentrating on what the other person is saying	Engaged, non-interruptive, feedback-oriented	Enhances understanding and trust, resolves misunderstandings	Improves negotiation outcomes, fosters collaboration

Assertive Communication	Expressing needs and opinions clearly and respectfully	Direct, clear, respectful	Promotes honest exchanges, prevents conflicts	Facilitates clear agreements, reduces ambiguity
Empathetic Communication	Showing understanding and compassion for others' perspectives	Reflective, validating, supportive	Builds rapport, strengthens relationships	Enhances conflict resolution, promotes team cohesion
Cross-Cultural Communication	Adapting communication styles to different cultural contexts	Flexible, respectful of cultural norms	Reduces misunderstandings, fosters inclusivity	Ensures smooth international collaboration, avoids cultural clashes

Negotiation and Dispute Resolution Techniques

Negotiation and dispute resolution are critical skills for quantity surveyors, as they often deal with conflicts arising from contract terms, project scope, or financial issues. Effective negotiation and dispute resolution contribute significantly to maintaining project harmony, ensuring fair outcomes, and avoiding costly delays or legal battles.

Negotiation involves discussions between parties aimed at reaching a mutually acceptable agreement. Successful negotiation requires preparation, clear communication, and an understanding of both parties' interests and objectives. Preparation includes gathering relevant information, understanding the other party's position, and defining clear goals for the negotiation. During the negotiation process, it is crucial to maintain a collaborative attitude rather than an adversarial one. This approach facilitates problem-solving and helps in finding creative solutions that satisfy all parties involved.

Effective negotiation also involves active listening and empathy. By genuinely understanding the other party's concerns and constraints, negotiators can identify common ground and address issues more constructively. It is important to articulate one's own position clearly and justify it with relevant data or evidence. Negotiators should also be prepared to make concessions where possible, while ensuring that such concessions do not undermine their key objectives.

Dispute resolution techniques come into play when negotiations fail to resolve conflicts. There are several methods for resolving disputes, including mediation, arbitration, and litigation. Mediation involves a neutral third party who facilitates communication and helps the parties reach a voluntary settlement. It is a non-binding process and is often used as a preliminary step before more formal dispute resolution methods. Arbitration, on the other hand, involves a neutral arbitrator who makes a binding decision based on the evidence and arguments presented by the parties. This method is more formal than mediation but less so than litigation. Litigation involves taking the dispute to court, where a judge or jury makes a binding decision. It is typically the last resort due to its time-consuming and costly nature.

The choice of dispute resolution method depends on various factors, including the complexity of the dispute, the relationship between the parties, and the contractual agreements in place. It is advisable for quantity surveyors to include dispute resolution clauses in contracts to outline the agreed-upon methods for handling potential conflicts. This proactive approach can help avoid misunderstandings and provide a clear path for resolving issues.

Real-World Examples of Successful Stakeholder Communication

Effective stakeholder communication is pivotal to the success of construction projects, influencing everything from project planning to execution and completion. Real-world examples of successful stakeholder communication provide valuable insights into how strategies and practices can be applied to achieve project goals and resolve issues efficiently.

One notable example is the Crossrail project in London,

one of Europe's largest infrastructure projects. The project involved multiple stakeholders, including government agencies, contractors, local communities, and the public. To manage these diverse interests, the project team implemented a comprehensive stakeholder engagement plan. This plan included regular updates through newsletters, public meetings, and dedicated communication channels. By actively engaging with stakeholders and addressing their concerns, the Crossrail project team was able to build trust and ensure smoother project delivery.

Another example is the Sydney Opera House, a landmark project that faced numerous communication challenges during its construction. The project's architect, Jørn Utzon, and the client, the New South Wales government, experienced significant communication issues, leading to delays and budget overruns. Despite these challenges, the resolution of disputes and the eventual completion of the project highlighted the importance of clear and continuous communication. The lessons learned from this project underscore the need for effective communication channels and stakeholder management strategies to navigate complex and high-profile projects successfully.

In the realm of private sector projects, the successful renovation of the Chrysler Building in New York City serves as an illustrative case. The renovation involved coordination between the building owner, tenants, contractors, and city officials. The project team employed a transparent communication approach, holding regular progress meetings and providing timely updates on construction activities. This proactive communication strategy helped mitigate potential disruptions and ensured that the renovation was completed on schedule while maintaining positive relationships with all stakeholders.

In each of these examples, effective communication was key to overcoming challenges and achieving project success. These cases demonstrate that clear, consistent, and transparent communication can help manage stakeholder expectations, address concerns promptly, and facilitate collaboration among various parties. Quantity surveyors can draw valuable lessons from these real-world examples to enhance their own communication practices, ensuring that they contribute positively to project outcomes and stakeholder satisfaction.

The Role of Soft Skills in Communication

In the field of quantity surveying, technical expertise is indispensable, but soft skills such as empathy and cultural awareness play a critical role in fostering effective communication and collaboration. These soft skills can significantly enhance interactions with stakeholders, contribute to conflict resolution, and support successful project outcomes.

Empathy involves understanding and being sensitive to the feelings, thoughts, and perspectives of others. For quantity surveyors, demonstrating empathy can lead to more productive interactions with clients, contractors, and other project stakeholders. For instance, when negotiating contracts or addressing disputes, an empathetic approach allows quantity surveyors to better appreciate the concerns and motivations of the other party. This understanding can lead to more amicable resolutions and foster stronger working relationships. Empathy also aids in recognizing the impact of project decisions on various stakeholders, which can guide more considerate and inclusive decision-making processes.

Cultural awareness is equally important, especially in projects involving diverse teams or international stakeholders. Cultural awareness involves recognizing and respecting cultural differences and adapting communication styles accordingly. For example, understanding cultural norms related to communication styles, decision-making processes, and negotiation tactics can help quantity surveyors navigate complex interactions more effectively. In multicultural project environments, being culturally aware can prevent misunderstandings and foster a more inclusive atmosphere. This includes being mindful of language barriers, different approaches to hierarchy and authority, and varying attitudes towards punctuality and deadlines.

Integrating these soft skills into professional practice involves both self-awareness and active engagement. Quantity surveyors should strive to cultivate empathy by actively listening to stakeholders, validating their concerns, and expressing genuine understanding. Similarly, developing cultural awareness requires ongoing education and exposure to diverse perspectives, as well as a willingness to adapt communication strategies to fit different cultural contexts.

While technical skills are essential for quantity surveyors, soft skills such as empathy and cultural awareness are crucial for effective communication and collaboration. These skills enhance the ability to manage stakeholder relationships, resolve conflicts, and ensure that projects progress smoothly. By prioritizing empathy and cultural sensitivity, quantity surveyors can contribute to more successful project outcomes and build stronger, more positive working relationships with all parties involved.

CHAPTER 12: COMMERCIAL AND CONTRACT MANAGEMENT SKILLS

Understanding Contract Law Principles

Contract law is a critical component of commercial and contract management, providing the legal framework within which contractual obligations and disputes are resolved. It governs the formation, execution, and enforcement of contracts and is essential for ensuring that agreements between parties are clear, enforceable, and fair.

At the core of contract law are several fundamental principles. The principle of offer and acceptance dictates that a contract is formed when one party makes an offer and the other party accepts it. The offer must be clear and unequivocal, and the acceptance must correspond exactly to the terms of the offer. Any deviation constitutes a counter-offer rather than acceptance.

The principle of consideration requires that each party to a contract provides something of value to the other. This value can be in the form of money, goods, services, or a promise to perform or refrain from performing a certain act. Consideration ensures that contracts are not mere promises but are based on mutual benefit.

Capacity refers to the legal ability of parties to enter into a contract. Generally, parties must be of legal age and possess the mental competency to understand the nature and consequences of their contractual obligations. Contracts involving parties who lack capacity, such as minors or individuals under duress, may be deemed void or voidable.

Legality of the contract's purpose is another fundamental principle. The subject matter of the contract must be legal and not against public policy. Contracts for illegal activities or those that contravene statutory regulations are unenforceable.

Mutual assent requires that both parties agree to the terms of the contract and intend to be bound by them. This assent is often demonstrated through written agreements, though oral contracts can also be valid provided they meet other contractual requirements.

Performance and breach are central to understanding how contracts are executed and enforced. Parties must perform their contractual obligations as agreed. Failure to do so constitutes a breach of contract, which can result in remedies such as damages, specific performance, or rescission. The non-breaching party may seek compensation or other legal remedies to address the breach.

Finally, dispute resolution mechanisms such as arbitration or mediation are often included in contracts to address any disagreements that arise. These mechanisms provide structured ways to resolve conflicts outside of traditional litigation, which can be time-consuming and costly.

Understanding these principles of contract law is essential

for effective commercial and contract management. They provide the foundation for drafting clear contracts, managing performance, and resolving disputes, ensuring that contractual relationships are managed efficiently and fairly.

Managing Claims and Variation Orders Effectively

Managing claims and variation orders is a vital aspect of contract management in the construction industry. It involves handling changes to the scope of work and addressing any claims for additional compensation or time extensions that may arise during the project. Effective management in this area helps ensure that projects remain on track, within budget, and compliant with contractual obligations.

Claims are requests made by one party to the contract for compensation or an extension of time due to changes or unforeseen circumstances. These claims often arise from variations in the work, delays caused by external factors, or disruptions that impact the project schedule. Proper management of claims involves several key steps:

- **Identification and Documentation**: The first step in managing claims is to identify and document the issue giving rise to the claim. This involves gathering evidence, including records of correspondence, site reports, and photographs, to support the claim. Accurate documentation is essential for substantiating the claim and ensuring that all relevant details are captured.

- **Assessment and Evaluation**: Once a claim is identified, it must be assessed to determine its validity and the impact on the project. This involves reviewing the contract terms, assessing the cause and effect of the issue, and evaluating the financial and time implications. It is crucial to ensure that the claim aligns with the contractual provisions for claims and variations.

- **Negotiation and Resolution**: After assessing the claim, the next step is to engage in negotiations with the other party to resolve. Effective negotiation involves clear communication, understanding the other party's perspective, and seeking a mutually acceptable solution. This may include agreeing on additional compensation, extending the project timeline, or other adjustments to the contract.

- **Documentation of Agreement**: Any agreement reached during negotiations should be documented formally. This includes drafting and signing a variation order or change order that outlines the agreed changes, the impact on the project, and any adjustments to the contract price or completion date. Proper documentation ensures that both parties have a clear understanding of the changes and their implications.

- **Implementation and Monitoring**: Once an agreement is documented, it must be implemented effectively. This involves updating project plans, schedules, and budgets to reflect the changes. Ongoing monitoring is essential to ensure that the changes are executed as agreed and that any further issues are promptly addressed.

Variation Orders are formal instructions issued by the client or project manager to change the scope of work. These changes can arise from design modifications, unforeseen conditions, or other factors affecting the project. Effective management of variation orders involves:

- **Request and Approval**: Variations should be formally requested and approved before work begins. The process typically involves submitting a variation request that outlines the proposed changes, the reasons for the variation, and any associated costs. Approval from the relevant parties is required to proceed with the changes.

- **Cost and Time Impact Assessment**: Each variation order should be assessed for its impact on the project cost and schedule. This involves calculating the additional costs associated with the variation and adjusting the project timeline accordingly. The assessment should be based on accurate and detailed information to ensure that the impacts are properly understood.

- **Contract Adjustment**: Once a variation order is approved, the contract should be adjusted to reflect the changes. This may involve amending the contract price, updating the project schedule, or making other necessary modifications. The revised contract should be documented and signed by all parties to formalize the changes.

- **Monitoring and Control**: After implementation, it is important to monitor the impact of the variation orders on the project. This includes tracking progress, controlling costs, and ensuring that the changes are delivered as specified. Effective control helps prevent additional issues and ensures that the project remains aligned with the revised plans.

Managing claims and variation orders effectively requires thorough documentation, careful assessment, clear communication, and diligent monitoring. By following these practices, project managers and quantity surveyors can address changes and claims efficiently, minimizing disruptions and maintaining project success.

Navigating Complex Contract Negotiations

Navigating complex contract negotiations is an essential skill for quantity surveyors and other professionals involved in construction projects. These negotiations are crucial for establishing clear, fair, and mutually beneficial agreements that set the foundation for successful project execution. Effective negotiation requires a thorough understanding of contract terms, the ability to address diverse stakeholder interests, and the skills to achieve a balanced outcome.

1. Preparation and Research

Successful contract negotiations begin with thorough preparation. This includes researching and understanding the key elements of the contract, the needs and objectives of all parties involved, and the broader context of the project. Preparation involves:

- **Understanding Contract Terms**: Familiarize yourself with standard contract terms and conditions, as well as any specific clauses relevant to the project. This includes provisions related to scope changes, performance requirements, and dispute resolution.

- **Stakeholder Analysis**: Identify and analyze the interests and priorities of all parties involved in the negotiation. Understanding their objectives and potential concerns helps in crafting proposals that address their needs while advancing your own goals.

- **Legal and Financial Considerations**: Assess any legal and financial implications of the contract terms. This may involve consulting with legal and financial experts to ensure that the contract complies with relevant regulations and is financially sound.

2. Setting Objectives and Strategy

Before entering negotiations, establish clear objectives and develop a strategy to achieve them. This involves:

- **Defining Goals**: Determine what you aim to achieve through the negotiation. This includes identifying non-negotiable terms and areas where flexibility is possible.

- **Developing a Strategy**: Create a negotiation strategy that outlines your approach, including how to address potential objections, manage conflicts, and make concessions if necessary. A well-defined strategy helps in maintaining focus and achieving desired outcomes.

3. Conducting Negotiations

During the negotiation process, effective communication and negotiation skills are critical. Key aspects include:

- **Effective Communication**: Clearly articulate your position and listen actively to the other party's concerns. Open and transparent communication helps build trust and facilitates constructive discussions.

- **Problem-Solving**: Address issues collaboratively and seek solutions that satisfy the interests of all parties. Focus on problem-solving rather than positions to find mutually acceptable resolutions.

- **Flexibility and Concessions**: Be prepared to make concessions and negotiate trade-offs to achieve a balanced agreement. Flexibility can help in finding common ground and advancing the negotiation.

4. Documenting Agreements

Once an agreement is reached, it is essential to document the terms accurately. This involves:

- **Drafting the Contract**: Prepare a comprehensive contract draft that reflects the agreed terms and conditions. Ensure that all key points discussed during negotiations are included and clearly defined.

- **Review and Approval**: Have the draft reviewed by all parties to ensure accuracy and completeness. Obtain formal approval and signatures from authorized representatives to finalize the contract.

5. Post-Negotiation Follow-Up

After the contract is signed, effective follow-up is necessary to ensure compliance and address any issues that arise. This includes:

- **Monitoring Compliance**: Track the implementation of contract terms and ensure that all parties adhere to their obligations. Regular monitoring helps in identifying and addressing potential issues early.

- **Managing Changes**: Address any changes or amendments to the contract that may be required. This involves negotiating and documenting any modifications in a formal manner.

Case Studies on Best Practices in Contract Management

Case studies on best practices in contract management provide valuable insights into how effective contract management can be achieved in various real-world scenarios. These case studies highlight successful strategies, common challenges, and lessons learned from different projects, offering practical examples for professionals to emulate. The examination of case studies allows for a deeper understanding of the nuances involved in contract management and the impact of best practices on project outcomes.

1. Case Study: Large-Scale Infrastructure Project

- **Project Overview**: This case study examines a major infrastructure project involving the construction of a new transportation hub. The project required extensive coordination between multiple contractors, government agencies, and stakeholders.

Best Practices

- **Clear Contractual Terms**: The project team established clear and detailed contractual terms that outlined the scope of work, performance requirements, and dispute resolution mechanisms. This helped prevent misunderstandings and disputes.

- **Stakeholder Engagement**: Regular meetings and consultations with stakeholders ensured that their concerns and interests were addressed throughout the project. This proactive approach facilitated smoother negotiations and approvals.

- **Risk Management**: The project included a comprehensive risk management plan that identified potential risks and outlined strategies for mitigation. This proactive approach minimized disruptions and ensured project continuity.

Lessons Learned

- **Importance of Detailed Contracts**: Clear and comprehensive contracts are crucial for managing complex projects and avoiding disputes. Ambiguities in contract terms can lead to misunderstandings and conflicts.

- **Effective Communication**: Regular and transparent communication with stakeholders is essential for managing expectations and addressing concerns promptly.

2. Case Study: Residential Development Project

- **Project Overview**: This case study focuses on a residential development project involving the construction of a mixed-use building with residential and commercial spaces. The project faced challenges related to budget overruns and scope changes.

Best Practices

- **Change Management Procedures**: The project implemented robust change management procedures to handle scope changes and budget adjustments. This included formal documentation of changes and approval processes.

- **Contractual Flexibility**: The contract included provisions for adjustments in case of unforeseen circumstances, allowing for flexibility in managing changes without significant delays.

- **Regular Performance Reviews**: Frequent performance reviews and progress assessments were conducted to ensure that the project stayed on track and any issues were addressed promptly.

Lessons Learned

- **Effective Change Management**: Establishing clear procedures for managing changes helps in controlling scope creep and budget overruns. Proper documentation and approval processes are essential for managing changes effectively.

- **Flexibility in Contracts**: Contracts should include provisions for flexibility to accommodate unforeseen changes and challenges without compromising project goals.

3. Case Study: International Construction Project

- **Project Overview**: This case study explores an international construction project involving the development of a commercial complex in a foreign country. The project faced challenges related to local regulations and cultural differences.

Best Practices

- **Local Compliance**: The project team conducted thorough research on local regulations and compliance requirements. Engaging local experts ensured that the project adhered to local laws and standards.
- **Cultural Sensitivity**: Understanding and respecting cultural differences helped in building strong relationships with local stakeholders and avoiding misunderstandings.

- **Detailed Contractual Agreements**: The contract included provisions to address potential legal and cultural issues, ensuring clarity and reducing the risk of disputes.

Lessons Learned

- **Adherence to Local Regulations**: Compliance with local regulations is crucial for the successful execution of international projects. Engaging local experts can help navigate complex legal and regulatory environments.

- **Cultural Awareness**: Understanding and respecting cultural differences can enhance collaboration and reduce the risk of conflicts in international projects.

4. Case Study: Renovation and Retrofit Project

- **Project Overview**: This case study examines a renovation and retrofit project for an existing commercial building. The project aimed to upgrade the building's facilities and improve energy efficiency.

Best Practices

- **Detailed Scope Definition**: The project team defined the scope of work in detail, including specific requirements for renovations and energy efficiency upgrades. This clarity helped in managing expectations and ensuring that the project met its goals.

- **Quality Control Measures**: Implementing rigorous quality control measures ensured that the renovations and retrofits were completed to high standards, meeting both regulatory and client expectations.

- **Stakeholder Involvement**: Engaging with stakeholders throughout the project, including tenants and facility managers, helped in addressing concerns and ensuring that the renovations met their needs.

Lessons Learned

- **Clear Scope Definition**: Clearly defining the scope of work is essential for managing renovation and retrofit projects effectively. This helps in avoiding scope creep and ensuring that the project objectives are met.

- **Quality Control**: Implementing stringent quality control measures is crucial for achieving high standards in renovation and retrofit projects.

Case studies on best practices in contract management provide valuable insights into effective strategies and common challenges. By examining real-world examples, quantity surveyors and construction professionals can gain practical knowledge and apply best practices to enhance their own contract management practices and achieve successful project outcomes.

CHAPTER 13:
PROFESSIONAL
PRACTICE AND ETHICS

Professional Responsibilities
of Quantity Surveyors

T he role of a Quantity Surveyor is integral to the construction industry, encompassing a wide array of responsibilities that ensure the effective and efficient management of construction projects. A Quantity Surveyor's professional duties are crucial at every stage of a project, from initial planning through to completion.

One of the primary responsibilities of a Quantity Surveyor is to manage project costs. This involves preparing detailed cost estimates, managing budgets, and ensuring that financial resources are used efficiently. The QS must accurately predict and control costs to avoid overspending and to deliver value for money. This responsibility extends to conducting cost analyses and offering advice on cost-effective measures and alternatives, thereby playing a significant role in financial planning and risk management.

In addition to cost management, Quantity Surveyors are tasked with the preparation of tender documentation and the assessment of bids from contractors. This includes ensuring that all contractual agreements are clear and comprehensive and that all terms and conditions are understood and agreed upon by all parties involved. The QS must also manage contract negotiations and facilitate communication between clients and contractors to resolve any issues that arise.

Quantity Surveyors are responsible for the measurement and valuation of work completed. They must accurately measure quantities, assess work progress, and prepare valuations for interim payments and final accounts. This responsibility is vital for ensuring that payments are made promptly and that the contractor is compensated fairly for work completed.

Moreover, a Quantity Surveyor must adhere to professional standards and ethical guidelines. They are expected to maintain high standards of integrity, transparency, and impartiality in their work. This involves avoiding conflicts of interest, providing honest and unbiased advice, and safeguarding the client's interests. Upholding these ethical principles is crucial for maintaining trust and credibility within the industry.

Quantity Surveyors also must stay abreast of industry developments, regulations, and best practices. This involves continuous professional development and keeping up-to-date with changes in legislation, technological advancements, and new methodologies in construction and cost management.

Furthermore, Quantity Surveyors are often involved in dispute resolution and claims management. They must be equipped to handle disputes between parties, interpret contract terms, and provide expert testimony if required. Effective dispute resolution requires strong negotiation skills and a deep understanding of both contractual obligations and industry practices.

Overall, the professional responsibilities of Quantity Surveyors are multifaceted and require a high level of expertise and professionalism. By fulfilling these responsibilities, QSs contribute significantly to the successful delivery of construction projects, ensuring that they are completed on time, within budget, and to the required quality standards.

Ethical Considerations in Quantity Surveying

Ethical considerations in Quantity Surveying are paramount to maintaining professionalism and trust within the construction industry. As Quantity Surveyors navigate complex projects and interact with various stakeholders, they are frequently faced with ethical dilemmas that require careful judgment and adherence to established ethical standards.

1. Integrity and Objectivity

A fundamental ethical principle for Quantity Surveyors is integrity. They are expected to be honest and transparent in their work, avoiding any form of misrepresentation or deceit. This involves providing accurate cost estimates and valuations, avoiding manipulation of data, and ensuring that all financial transactions are conducted with honesty. Objectivity is closely related to integrity; Quantity Surveyors must provide impartial advice, free from personal bias or external pressures. Their judgments and recommendations should be based solely on facts and professional expertise.

2. Confidentiality

Confidentiality is a critical ethical consideration in Quantity Surveying. Quantity Surveyors often handle sensitive information, including financial data, contract details, and proprietary project information. They are obligated to protect this information from unauthorized disclosure and ensure that it is used solely for its intended purpose. Breaching confidentiality can damage professional relationships and compromise the integrity of the project.

3. Professional Competence

Maintaining professional competence is essential for ethical practice. Quantity Surveyors must ensure that they possess the necessary skills, knowledge, and qualifications to perform their duties effectively. This includes staying current with industry developments, changes in legislation, and advancements in technology. Continuous professional development and training are vital to uphold the standards of competence and to provide informed and accurate advice.

4. Avoiding Conflicts of Interest

Quantity Surveyors must be vigilant in avoiding conflicts of interest. A conflict of interest arises when personal interests or relationships compromise the ability to act in the best interests of the client or project. Quantity Surveyors should disclose any potential conflicts and take appropriate measures to manage or mitigate them. This may involve recusing themselves from certain decisions or seeking guidance from professional bodies.

5. Compliance with Legal and Professional Standards

Adhering to legal and professional standards is a key aspect of ethical practice. Quantity Surveyors must comply with relevant laws, regulations, and codes of conduct. This includes understanding and applying contract law, health and safety regulations, and industry best practices. Compliance ensures that their work meets legal requirements and upholds the standards expected by the profession.

6. Accountability and Transparency

Quantity Surveyors are accountable for their actions and decisions. They must be transparent in their reporting, provide clear and detailed documentation, and be willing to explain and justify their work to clients and other stakeholders. Accountability ensures that their work can be audited and reviewed, maintaining trust and credibility within the industry.

7. Respect and Fair Treatment

Respect and fairness are essential ethical considerations. Quantity Surveyors should treat all stakeholders with respect and ensure that their dealings are fair and equitable. This involves avoiding discriminatory practices, treating all parties with dignity, and addressing any grievances or concerns respectfully. Table 20 outlines the key ethical principles and corresponding considerations that quantity surveyors should adhere to in their professional practice.

Table 20: Ethical Principles and Considerations for Quantity Surveyors

Ethical Principle	Consideration
Integrity	Always act with honesty and fairness. Avoid conflicts of interest and disclose any potential biases.
Confidentiality	Respect the confidentiality of information received in the course of professional duties.
Transparency	Provide clear, accurate, and timely information to clients and stakeholders. Avoid misleading information.
Competence	Ensure that services are provided competently, with the necessary skills and knowledge.
Accountability	Take responsibility for actions and decisions. Be prepared to explain and justify them.
Professionalism	Maintain a high standard of professional behavior and appearance. Comply with relevant laws and regulations.
Respect	Treat colleagues, clients, and other stakeholders with respect and courtesy.
Environmental Responsibility	Consider the environmental impact of construction projects and promote sustainable practices.
Continuous Learning	Commit to lifelong learning and staying

	updated with industry developments and best practices.

Legal Frameworks and Compliance

Legal frameworks and compliance are essential components of Quantity Surveying, providing the regulatory and legal context within which Quantity Surveyors operate. Understanding these frameworks ensures that projects are executed in accordance with relevant laws, regulations, and standards, safeguarding the interests of clients, stakeholders, and the public.

1. Understanding Contract Law

Contract law forms the foundation of construction projects, governing the agreements between parties involved, including clients, contractors, and subcontractors. Quantity Surveyors must be well-versed in contract law principles, including the formation, execution, and termination of contracts. This knowledge is critical in drafting clear and enforceable contract terms, interpreting contractual obligations, and managing disputes that may arise.

2. Compliance with Building Regulations

Building regulations set the standards for the construction, design, and maintenance of buildings to ensure safety, health, and sustainability. Quantity Surveyors must ensure that all aspects of a project comply with these regulations, which may include structural integrity, fire safety, accessibility, and environmental considerations. Non-compliance can result in legal penalties, delays, and increased costs, making adherence to these regulations a priority.

3. Health and Safety Regulations

Health and safety are paramount in construction projects,

where the risk of accidents and injuries is significant. Quantity Surveyors must be knowledgeable about health and safety regulations, including those related to construction site practices, equipment usage, and worker protection. They play a role in ensuring that safety measures are incorporated into project planning and execution, conducting risk assessments, and monitoring compliance throughout the project lifecycle.

4. Environmental Regulations and Sustainability

Environmental regulations have become increasingly important as the construction industry seeks to minimize its environmental impact. Quantity Surveyors must be aware of regulations related to environmental protection, waste management, and energy efficiency. This includes compliance with laws governing emissions, hazardous materials, and resource conservation. In addition, Quantity Surveyors may be involved in achieving sustainability certifications, such as LEED or BREEAM, which require adherence to specific environmental standards.

5. Professional Standards and Codes of Conduct

Quantity Surveyors are often governed by professional bodies that set standards for ethical and professional conduct. These bodies, such as the Royal Institution of Chartered Surveyors or the South African Council for the Quantity Surveying Profession, provide guidelines on ethical behavior, competence, and accountability. Compliance with these standards is essential for maintaining professional accreditation and credibility. Quantity Surveyors must adhere to these codes of conduct, which outline expectations regarding integrity, objectivity, confidentiality, and competence.

6. Intellectual Property and Data Protection

In an era of digitalization, Quantity Surveyors must also navigate issues related to intellectual property and data protection. This includes understanding the ownership and use of digital data, respecting copyright laws, and ensuring the confidentiality and security of client information. Compliance with data protection regulations, such as the General Data Protection Regulation in the European Union, is crucial in managing personal and sensitive data responsibly.

7. Dispute Resolution and Legal Processes

Dispute resolution is a critical aspect of legal compliance in construction projects. Quantity Surveyors must be familiar with various dispute resolution mechanisms, including negotiation, mediation, adjudication, and arbitration. Understanding these processes enables Quantity Surveyors to advise on the most appropriate course of action and to manage disputes efficiently and effectively, minimizing disruption to the project.

The Role of Professional Bodies and Associations

Professional bodies and associations play a crucial role in the field of Quantity Surveying, providing governance, standards, and support for professionals in the industry. These organizations not only uphold the integrity and quality of the profession but also offer a range of benefits to their members, including professional development, networking opportunities, and advocacy.

1. Setting Standards and Best Practices

Professional bodies such as the Royal Institution of Chartered Surveyors and the South African Council for the Quantity Surveying Profession set standards and best practices for Quantity Surveyors. These standards cover various aspects of the profession, including ethical conduct, technical competence, and professional behavior. By establishing these benchmarks, professional bodies ensure that Quantity Surveyors operate to a consistent and high standard, providing reliable and high-quality services to clients.

2. Accreditation and Certification

Accreditation and certification are key functions of professional bodies. They assess and certify the qualifications and competencies of Quantity Surveyors, granting them professional titles such as Chartered Quantity Surveyor (CQS) or Professional Quantity Surveyor (PrQS). This certification process typically involves rigorous examinations, assessments, and interviews, ensuring that only individuals with the necessary knowledge, skills, and experience receive professional recognition. Accredited Quantity Surveyors benefit from increased credibility and marketability, as these certifications are widely recognized and respected in the industry.

3. Continuing Professional Development

Continuing Professional Development (CPD) is essential for maintaining and enhancing the skills and knowledge of Quantity Surveyors. Professional bodies often mandate a certain amount of CPD hours annually, which can be achieved through various activities such as attending workshops, seminars, conferences, or completing online courses. These bodies provide resources and opportunities for CPD, ensuring that members stay current with industry developments, new technologies, and regulatory changes. CPD is crucial for professional growth and for keeping up with the evolving demands of the construction industry.

4. Advocacy and Representation

Professional bodies advocate for the interests of Quantity Surveyors and the construction industry as a whole. They engage with government bodies, regulators, and other stakeholders to influence policies, regulations, and legislation that impact the profession. This advocacy work helps to create a favorable operating environment for Quantity Surveyors, addressing issues such as industry standards, safety regulations, and environmental concerns. Additionally, these organizations represent the profession in public forums, promoting the value and importance of Quantity Surveying to the broader community.

5. Ethical Guidance and Disciplinary Procedures

Maintaining ethical standards is a core function of professional bodies. They provide ethical guidelines and codes of conduct that members must adhere to, ensuring integrity, transparency, and accountability in their professional practice. These guidelines cover various areas, including conflicts of interest, confidentiality, and professional relationships. In cases of misconduct or unethical behavior, professional bodies have established disciplinary procedures to investigate complaints and impose sanctions if necessary. This oversight helps to protect the reputation of the profession and maintain public trust.

6. Networking and Collaboration

Professional bodies offer numerous opportunities for networking and collaboration among Quantity Surveyors and other construction professionals. Events such as conferences, seminars, and workshops bring together experts from various fields, facilitating the exchange of knowledge, ideas, and best practices. These gatherings also provide a platform for members to build professional relationships, collaborate on projects, and explore new business opportunities. Networking is particularly valuable for career development, as it can lead to mentorship, partnerships, and job opportunities.

7. Research and Knowledge Dissemination

Research and knowledge dissemination are vital activities of professional bodies. They support and conduct research on topics relevant to Quantity Surveying and the construction industry, contributing to the advancement of the profession. This research is often published in journals, reports, and other publications, providing members with access to the latest findings and trends. Additionally, professional bodies may offer training and resources on new methodologies, technologies, and practices, helping Quantity Surveyors stay informed and competitive.

Continuing Professional Development for Quantity Surveyors

Continuing Professional Development is an essential aspect of a Quantity Surveyor's career, aimed at maintaining and enhancing professional competence, skills, and knowledge throughout their career. CPD is not just a requirement for maintaining professional registration or accreditation; it is also crucial for adapting to the evolving demands of the construction industry, which is constantly influenced by new technologies, regulations, and best practices.

1. Importance of CPD

The construction industry is characterized by rapid changes and advancements, including the introduction of new building materials, technologies, regulatory requirements, and management practices. CPD ensures that Quantity Surveyors remain up-to-date with these changes, thereby maintaining their professional relevance and ability to provide high-quality services. It also fosters a culture of lifelong learning, encouraging professionals to continually seek improvement and innovation in their practice.

2. CPD Requirements and Guidelines

Professional bodies such as the Royal Institution of Chartered Surveyors, the South African Council for the Quantity Surveying Profession (SACQSP), and other relevant associations typically establish specific CPD requirements for their members. These requirements often include a minimum number of CPD hours or units that must be completed within a given period, usually annually or biannually. CPD activities can vary widely, including formal education, training courses, seminars, workshops, conferences, webinars, and even self-directed learning such as reading professional literature or conducting research.

3. Types of CPD Activities

CPD activities can be broadly categorized into structured and unstructured activities:

- **Structured CPD**: This includes formal learning activities such as attending accredited courses, workshops, seminars, or conferences. These activities are often pre-approved by professional bodies and provide clear learning outcomes. Structured CPD is typically more focused and provides specific skills or knowledge relevant to the practice of Quantity Surveying.

- **Unstructured CPD**: This includes informal learning activities such as self-study, reading industry journals or technical papers, participating in professional discussions or forums, and on-the-job learning. While unstructured CPD is more flexible and can be tailored to individual interests and needs, it also requires self-discipline to ensure that the learning objectives are met.

4. Planning and Recording CPD

Effective CPD requires careful planning and systematic recording. Quantity Surveyors should assess their current skills and knowledge, identify areas for improvement, and set specific learning objectives. This process involves creating a CPD plan that outlines the activities to be undertaken, the resources needed, and the expected outcomes. Recording CPD activities is equally important, as it provides evidence of professional development and helps track progress over time. Professional bodies often provide templates or online platforms for recording CPD, which members can use to log their activities and reflect on their learning experiences.

5. Benefits of CPD

CPD offers numerous benefits to Quantity Surveyors, including:

- **Enhanced Competence and Confidence**: By keeping up-to-date with the latest industry developments, Quantity Surveyors can improve their technical skills and knowledge, leading to greater confidence in their professional capabilities.

- **Career Advancement**: CPD can open up new career opportunities by equipping Quantity Surveyors with the skills needed for higher-level roles or specializations. It can also enhance employability by demonstrating a commitment to professional growth and excellence.

- **Professional Recognition and Compliance**: Many professional bodies require CPD as a condition for maintaining membership or certification. Completing CPD requirements ensures compliance with these standards and can lead to recognition through awards or designations.

- **Networking and Professional Relationships**: Participating in CPD activities, such as conferences and workshops, provides opportunities to network with peers, experts, and industry leaders. These connections can lead to collaborative projects, mentorship, and career development opportunities.

6. Challenges and Solutions

Despite its importance, CPD can present challenges, including time constraints, financial costs, and limited access to resources or opportunities. Quantity Surveyors can overcome these challenges by:

- **Prioritizing CPD**: Setting clear priorities and allocating time for CPD activities can help manage time constraints. Employers can support this by allowing time off for professional development or incorporating CPD into performance reviews.

- **Seeking Cost-Effective Options**: Many CPD activities, such as webinars, online courses, and professional literature, are available at low or no cost. Quantity Surveyors can also look for funding opportunities or employer-sponsored training.

- **Leveraging Technology**: Online platforms and e-learning tools offer flexible and accessible CPD options. These resources allow Quantity Surveyors to learn at their own pace and from any location, making it easier to fit CPD into a busy schedule.

Continuing Professional Development is a vital component of a Quantity Surveyor's professional journey. It ensures that they remain competent and competitive in a dynamic industry, enhances their career prospects, and upholds the standards and reputation of the profession. By engaging in a variety of

CPD activities, Quantity Surveyors can continually grow their expertise, adapt to new challenges, and contribute positively to the construction industry.

CONCLUSION

Key Takeaways from Each
Part of the Book

T he book provides a comprehensive understanding of the multifaceted role of quantity surveying in the construction sector. Each part of the book has been designed to prepare both aspiring and experienced quantity surveyors with the knowledge and skills necessary to excel in their careers.

Part 1: Introduction to
Quantity Surveying

The introductory chapters lay the foundation for understanding the role of a quantity surveyor, exploring various career paths, and highlighting essential tools and resources. It emphasizes the importance of foundational knowledge in construction contracts, risk management, and the pivotal role quantity surveyors play in ensuring project efficiency and cost-effectiveness.

Part 2: Mastering Measurement
and Estimating

This section delves into the core technical skills required for measurement and estimating. Readers learn about standard measurement techniques, cost estimating methods, and the preparation of detailed bills of quantities. The focus on practical skills and the use of digital tools underscores the importance of accuracy and efficiency in cost management.

Part 3: Quantity Surveying Throughout the Project Lifecycle

Here, the book covers the involvement of quantity surveyors at different stages of the construction project lifecycle, from pre-construction cost planning and feasibility studies to managing variations and final account settlement. The emphasis on best practices and case studies provides real-world insights into the challenges and solutions encountered in practice.

Part 4: Essential Skills for Quantity Surveyors

This part highlights the critical soft skills needed for effective communication, negotiation, and stakeholder management. It underscores the significance of empathy, cultural awareness, and professionalism in fostering positive working relationships and resolving disputes.

Part 5: Emerging Trends and Future Directions

The final chapters explore the evolving landscape of quantity surveying, including sustainable construction practices, digital technologies like Building Information Modeling, and the importance of ethics and professional development. This section prepares readers for the future challenges and opportunities in the industry, encouraging a forward-thinking approach.

Encouraging Continuous Learning and Professional Development

The construction industry is continually evolving, driven by technological advancements, changing regulations, and new sustainability standards. As such, continuous learning and professional development are crucial for quantity surveyors to remain relevant and competitive. This book emphasizes the value of lifelong learning through Continuing Professional Development, encouraging readers to seek out new knowledge, skills, and certifications.

The emphasis on CPD underscores the importance of staying abreast of industry trends, enhancing technical expertise, and developing soft skills. Whether through formal education, attending seminars, engaging with professional bodies, or self-directed study, CPD is portrayed as an integral part of a quantity surveyor's career progression.

Preparing for Future Challenges and Opportunities in Quantity Surveying

As the construction industry faces increasing demands for sustainability, efficiency, and innovation, quantity surveyors must be prepared to navigate these challenges. The book highlights several key areas where future challenges and opportunities may arise:

- **Sustainable Practices**: With growing emphasis on green building certifications and sustainable construction practices, quantity surveyors must develop expertise in cost-benefit analysis and the assessment of environmental impacts.

- **Digital Transformation**: The integration of technologies like BIM, digital measurement tools, and data analytics is transforming the industry. Quantity surveyors need to adapt to these changes by developing proficiency in these tools and understanding their applications in cost management and project coordination.

- **Globalization and Cultural Awareness**: As projects become increasingly global, the ability to work across different cultures and regulatory environments becomes essential. Quantity surveyors must be equipped to manage international projects, understand diverse market conditions, and navigate cross-cultural communication challenges.

- **Ethics and Professionalism**: Maintaining high ethical standards is crucial in all professional practices. The book underscores the importance of ethical decision-making, compliance with legal frameworks, and the role of professional bodies in upholding these standards.

This book serves as a comprehensive guide for those entering the field and seasoned professionals seeking to enhance their skills. It encourages a proactive approach to learning and development, preparing quantity surveyors to meet the evolving demands of the industry and seize emerging opportunities. The emphasis on practical skills, professional ethics, and continuous improvement positions readers for success in a dynamic and challenging profession.

ACKNOWLEDGEMENT

I would like to express my deepest gratitude to all those who contributed to the creation of this book. To my colleagues and peers in the construction and quantity surveying fields, your insights and shared experiences have enriched the content and perspective of this work.

A special thank you to my academic mentors and instructors, whose guidance has been instrumental in shaping my understanding and passion for this profession. Your teachings have been invaluable, and your encouragement has fueled my commitment to sharing knowledge with others.

I am also profoundly grateful to my family and friends for their unwavering support, patience, and belief in my vision. Your love and encouragement have been a constant source of strength and motivation throughout this journey.

Finally, to the readers and fellow professionals, your dedication to learning and professional growth is the true inspiration behind this book. I hope that this work serves as a useful resource in your career and contributes positively to the field of quantity surveying.

ABOUT THE AUTHOR

Steven Smith, Ph.d.

Steven Smith is a renowned expert in the field of Construction Management, with a wealth of knowledge in academia. Holding a doctorate in Construction Management, Steven has dedicated his career to advancing the field and contributing to its body of knowledge.

Throughout his academic journey, Steven's passion for understanding the intricacies of construction processes and finding innovative solutions to industry challenges became evident. His doctoral research focused on optimizing project management practices and enhancing productivity in construction projects, leading to a profound understanding of various aspects of construction management and their impact on project success.

BOOKS BY THIS AUTHOR

The Essential Dictionary For Quantity Surveyors: An Indispensable Resource For Students, Professionals, And Academics

Master the Language of Quantity Surveying

Navigating the complex world of construction requires a deep understanding of specialized terminology. Quantity surveyors, the financial stewards of construction projects, rely on a precise vocabulary to effectively communicate, collaborate, and make informed decisions. The Essential Dictionary for Quantity Surveyors is your indispensable guide to mastering this intricate language.

Why You Need This Book:

Clear and Concise Definitions: Gain a comprehensive understanding of essential Quantity Surveying terms.

Real-world Context: Apply your knowledge through practical examples and illustrations.

Stay Ahead of the Curve: Keep up with the latest industry trends and terminology.

Enhance Communication: Improve collaboration with clients, contractors, and other stakeholders.

Boost Your Career: Demonstrate your expertise and elevate your professional standing.

From fundamental concepts to advanced practices, this dictionary covers a wide range of topics, including measurement, valuation, contract administration, and sustainability. Whether you're a student, a seasoned professional, or an academic, this book will be your go-to reference for accurate and up-to-date information.

Don't let complex terminology hinder your success. Invest in your professional growth with The Essential Dictionary for Quantity Surveyors.

The Fundamentals Of Construction Quantity Surveying

The Fundamentals of Construction Quantity Surveying

Quantity surveying is an important profession in the construction industry, and quantity surveyors are in high demand. If you're interested in a rewarding construction career with excellent job prospects, then quantity surveying is the field for you.

This book covers everything you need to know to get started in the profession, from the basics of cost estimation and measurement to contract administration and dispute resolution.

The author employs a clear, concise, and easy-to-understand writing style to convey the book's content. It is packed with practical examples to provide extensive perspectives on different subjects.

The book is the ideal resource for anyone who wants to:

Learn more about the quantity surveying profession
Get a job in quantity surveying
Advance their career in quantity surveying
Start their own quantity surveying business
By reading this book, you'll learn the essential skills and knowledge you need to become a successful quantity surveyor, including:
How to measure and estimate the cost of construction work
How to manage construction contracts
How to resolve disputes
How to use the latest technology and software
How to build a successful career in quantity surveying

If you aspire to have a successful career in quantity surveying, I recommend ordering your copy of this book today!